JOURNAL OF APPLIED LOGICS - IFCOLOG
JOURNAL OF LOGICS AND THEIR APPLICATIONS

Volume 5, Number 3

June 2018

Disclaimer

Statements of fact and opinion in the articles in Journal of Applied Logics - IfCoLog Journal of Logics and their Applications (JAL-FLAP) are those of the respective authors and contributors and not of the JAL-FLAP. Neither College Publications nor the JAL-FLAP make any representation, express or implied, in respect of the accuracy of the material in this journal and cannot accept any legal responsibility or liability for any errors or omissions that may be made. The reader should make his/her own evaluation as to the appropriateness or otherwise of any experimental technique described.

ISBN 978-1-84890-279-4
ISSN (E) 2055-3714
ISSN (P) 2055-3706

College Publications
Scientific Director: Dov Gabbay
Managing Director: Jane Spurr

http://www.collegepublications.co.uk

Printed by Lightning Source, Milton Keynes, UK

Modal and Temporal Logic
Carlos Areces
Melvin Fitting
Victor Marek
Mark Reynolds.
Frank Wolter
Michael Zakharyaschev

Automated Inference Systems and Model Checking
Ed Clarke
Ulrich Furbach
Hans Juergen Ohlbach
Volker Sorge
Andrei Voronkov
Toby Walsh

Formal Methods: Specification and Verification
Howard Barringer
David Basin
Dines Bjorner
Kokichi Futatsugi
Yuri Gurevich

Logic and Software Engineering
Manfred Broy
John Fitzgerald
Kung-Kiu Lau
Tom Maibaum
German Puebla

Logic and Constraint Logic Programming
Manuel Hermenegildo
Antonis Kakas
Francesca Rossi
Gert Smolka

Logic and Databases
Jan Chomicki
Enrico Franconi
Georg Gottlob
Leonid Libkin
Franz Wotawa

Logic and Physics (space time. relativity and quantum theory)
Hajnal Andreka
Kurt Engesser
Daniel Lehmann
lstvan Nemeti
Victor Pambuccian

Logic for Knowledge Representation and the Semantic Web
Franz Baader
Anthony Cohn
Pat Hayes
Ian Horrocks
Maurizio Lenzerini
Bernhard Nebel

Tactical Theorem Proving and Proof Planning
Alan Bundy
Amy Felty
Jacques Fleuriot
Dieter Hutter
Manfred Kerber
Christoph Kreitz

Logic and Algebraic Programming
Jan Bergstra
John Tucker

Logic in Mechanical and Electrical Engineering
Rudolf Kruse
Ebrahaim Mamdani

Logic and Law
Jose Carmo
Lars Lindahl
Marek Sergot

Applied Non-classical Logic
Luis Farinas del Cerro
Nicola Olivetti

Mathematical Logic
Wilfrid Hodges
Janos Makowsky

Cognitive Robotics: Actions and Causation
Gerhard Lakemeyer
Michael Thielscher

Type Theory for Theorem Proving Systems
Peter Andrews
Chris Benzmüller
Chad Brown
Dale Miller
Carsten Schlirmann

Logic Applied in Mathematics (including e-Learning Tools for Mathematics and Logic)
Bruno Buchberger
Fairouz Kamareddine
Michael Kohlhase

Logic and Computational Models of Scientific Reasoning
Lorenzo Magnani
Luis Moniz Pereira
Paul Thagard

Logic and Multi-Agent Systems
Michael Fisher
Nick Jennings
Mike Wooldridge

Logic and Neural Networks
Artur d'Avila Garcez
Steffen Holldobler
John G. Taylor

Logic and Planning
Susanne Biundo
Patrick Doherty
Henry Kautz
Paolo Traverso

Algebraic Methods in Logic
Miklos Ferenczi
Rob Goldblatt
Robin Hirsch
Idiko Sain

Non-monotonic Logics and Logics of Change
Jurgen Dix
Vladimir Lifschitz
Donald Nute
David Pearce

Logic and Learning
Luc de Raedt
John Lloyd
Steven Muggleton

Logic and Natural Language Processing
Wojciech Buszkowski
Hans Kamp
Marcus Kracht
Johanna Moore
Michael Moortgat
Manfred Pinkal
Hans Uszkoreit

Fuzzy Logic Uncertainty and Probability
Didier Dubois
Petr Hajek
Jeff Paris
Henri Prade
George Metcalfe
Jon Williamson

SCOPE AND SUBMISSIONS

This journal considers submission in all areas of pure and applied logic, including:

pure logical systems
proof theory
constructive logic
categorical logic
modal and temporal logic
model theory
recursion theory
type theory
nominal theory
nonclassical logics
nonmonotonic logic
numerical and uncertainty reasoning
logic and AI
foundations of logic programming
belief revision
systems of knowledge and belief
logics and semantics of programming
specification and verification
agent theory
databases

dynamic logic
quantum logic
algebraic logic
logic and cognition
probabilistic logic
logic and networks
neuro-logical systems
complexity
argumentation theory
logic and computation
logic and language
logic engineering
knowledge-based systems
automated reasoning
knowledge representation
logic in hardware and VLSI
natural language
concurrent computation
planning

This journal will also consider papers on the application of logic in other subject areas: philosophy, cognitive science, physics etc. provided they have some formal content.

Submissions should be sent to Jane Spurr (jane.spurr@kcl.ac.uk) as a pdf file, preferably compiled in LaTeX using the IFCoLog class file.

CONTENTS

ARTICLES

Argument Strength in Formal Argumentation

Mathieu Beirlaen*
Ghent University, Belgium
mathieubeirlaen@gmail.com

Jesse Heyninck, Pere Pardo, Christian Strasser[†]
Ruhr-Universität Bochum, Germany
{jesse.heyninck,pere.pardoventura,christian.strasser}@rub.de

Some arguments are stronger than others. What makes an argument strong may vary with the reasoning context. In epistemic contexts, where the main interest is a true account of a given subject matter, a strong argument is based on reliable information and reliable (not necessarily deductive) inferences from this information. In practical decision making, we may care more about arguments promoting certain specific values than about arguments promoting other values. In didactic contexts we may be interested mainly in the explanatory strength of an argument, and possibly in the extent to which it is understood by the target audience. Turning our focus to the social dimension of argumentation, we may consider the strength of an argument relative to its dialectic nature. Will it refute many arguments the opponents bring forward? Will it make one's overall stance more vulnerable, etc.?

Computational argumentation provides a formal theory of argument construction and evaluation. Given the prominent place the notion of argument strength has in discursive contexts, it is key for this research program to develop an account of it. In this article we highlight some of the main conceptual developments in this domain on the notion of argument strength. In computational argumentation we can distinguish between three tiers or dimensions of argumentation. The notion

This special issue came about in the wake of the first *Argument Strength* workshop, held at the Ruhr-University of Bochum from November 30th till December 2nd of 2016. We thank all participants for their contributions at the workshop, and for valuable discussions which helped to shape this overview. We also thank the referees of papers submitted to this volume for their generous help and expertise.

*This research was funded by the Flemish Research Foundation (FWO-Vlaanderen).

†Research for this article was sponsored by a Sofja Kovalevskaja award of the Alexander von Humboldt Foundation, funded by the German Ministry for Education and Research.

of argument strength comes into play in each of these. We will use these tiers to structure and guide the discussion below by dedicating one section of this article to each tier:

Sec. 1 The *support dimension*. Which premises does the argument rely on? How reliable are they? To what extent do they support the conclusion? Which inference rules were used in the argument's construction on the basis of these premises, and how reliable are these rules? These questions concern an argument's construction and degree of support independently of the presence of other arguments. In informal logic, the support dimension corresponds to the *illative core* of an argument: the argument seen as a premise-conclusion structure which provides for the argument's conclusion [32, 93].

Sec. 2 The *dialectic dimension*. We do not assess arguments in isolation. A good assessment takes into account alternative points of view and possible objections. For this, we need to consider counter-arguments. Johnson refers to this dimension as the *dialectical tier* of argumentation [93]. In a formal account of argumentation, this dialectics is mainly represented by relations of argumentative attack and defeat between arguments.

Sec. 3 The *evaluative dimension*. There exist various ways of determining which arguments in a given set can be considered justified or unjustified. The choice of a particular method is a highly context-dependent matter. In practical decision making, we may attach more importance to a specific value promoted by some argument. In certain epistemic contexts we may want to adopt a rather skeptical stance on argument acceptance, while in others we may want to be more credulous, and so on. Given a set of arguments and a relation of argumentative defeat over this set, the evaluative dimension consists of a large variety of 'semantics' for determining the acceptability status of arguments and their respective claims relative to the context at hand.

The current main point of reference in formal argumentation is Dung's approach from [63]. Dung defines an *argumentation framework* as a pair consisting of a set of arguments and a relation of argumentative attack over arguments. The second member of this pair belongs to the dialectic tier (Section 2), while the first relates to the support tier (Section 1). In the support tier we zoom in on the illative core or internal *structure* of arguments. The subfield of formal argumentation that takes this dimension into account is called *structured argumentation* [25]. Given an argumentation framework, Dung provides various *semantics* for determining the acceptability status of an argument in the framework. These belong to the evaluative tier. We will have more to say about these in Section 3. We end with a short overview of different approaches to argument accrual, a topic that connects to all three tiers of computational argumentation (Section 4), and with a discussion on the power of formal approaches to argumentation to represent and incorporate existing work in the field of non-monotonic logic (Section 5).

1 The Support Dimension: Argument Construction

Systems of structured or instantiated argumentation specify how arguments are constructed relative to a *premise set* and a number of *inference rules*. Premises are formulas in a given formal language. They represent the evidence or information on the basis of which we build arguments. Besides premises, we have rules at our disposal for inferring new formulas from others. Arguments are considered the result of applying inference rules to the given premises and, possibly, of chaining such applications. In applications of automated reasoning a given set of premises and a given set of inference rules form a *knowledge base* from which arguments are generated. Such a knowledge base can be used to model, for example, legal argumentation (see [19] for an overview), reasoning with conditional obligations [18, 103], default reasoning [160, 159], logic programming with negation as failure [52], autoepistemic reasoning [37], causal inference [34] or reasoning with plausible assumptions [37].

In Section 1.1 below we present some standard accounts of argument construction in the subfield of structured argumentation. The support provided by an argument for its conclusion is determined by the degree of support of its premises, and by the degree of support provided by the inference rules applied in its construction. We discuss these determinants in Sections 1.2 and 1.3 respectively. Degrees of support are relative to a pre-given ordering over the premises and/or rules in the knowledge base. Differences in the type of ordering chosen may have important consequences for the ensuing account of argument strength. These differences are discussed in Section 1.4.

1.1 A basic matrix of argument representation

A first source of variation with respect to the way arguments are constructed, concerns the expressive power of the system at hand. The formal language in which premise sets are encoded may or may not contain disjunctions, predicates, etc. Moreover, systems may or may not allow for a distinction between strict and defeasible rules. *Strict* rules are deductive and non-retractable, while *defeasible* rules can be retracted and their consequents do not follow with absolute certainty from the fact that the antecedent holds. A similar distinction in terms of certainty/retractability

System	StrPrem	DefPrem	StrRules	DefRules
ABA	✓	✓	✓	×
ASPIC⁺	✓	✓	✓	✓
DA	×	✓	✓	×
DeLP	✓*	✓*	✓	✓
DL	✓	✓*	✓	✓
SBA	×	✓	✓	×

Table 1

can be made between *strict premisses* (facts) and *defeasible premisses* (assumptions). Strict rules often correspond to valid deductions in some core logic, such as classical logic (**CL**), intuitionistic logic, or some modal or paraconsistent logic. Systems of structured argumentation are usually quite flexible when it comes to the choice of this deductive core. In Table 1 we provide an overview of some central systems of structured argumentation and their capacities to represent different types of rules and premisses.

'StrPrem' and 'StrRules' refer to the capacity to represent strict premisses and strict rules respectively. Similarly, 'DefPrem' and 'DefRules' refer to the capacity to represent defeasible premisses and defeasible rules respectively. A ✓*-symbol denotes systems that can only express premisses indirectly via the use of rules, e.g. as rules with an empty body $\top \Rightarrow \varphi$ (in DeLP and DL).[1]

ABA is the framework of assumption-based argumentation by Bondarenko et al. [37]; ASPIC⁺ is the framework for structured argumentation developed by Prakken & Modgil [112, 113]; DA is Besnard & Hunter's framework of deductive argumentation [28]; DeLP is the defeasible logic programming framework by García & Simari [72]; DL is Nute's defeasible logic [119]; and SBA is the sequent-based argumentation framework as found in Arieli & Straßer [9].

Clearly, defeasibility comes in various flavours. In ABA, for instance, all inference rules are strict but premisses can be defeasible. This type of defeasible reasoning has been labelled plausible reasoning by Rescher [116]. In ASPIC⁺, both premisses and inference rules may be defeasible. Defeasible inference rules may represent generalizations that allow for exceptions, such as the default rule "birds typically

---

[1] Although DA and SBA do not accommodate strict premisses in their standard formulations, one can model these e.g. by means of adding axioms to the core logic. For instance, in DA strict premisses ϕ can then be introduced by arguments of the form $(\emptyset, \phi)$. For paradigmatic core logics, such as **CL**, these arguments cannot be attacked, because any rebuttal-attacker $(\Gamma, \neg\phi)$ would be such that Γ is inconsistent in the enriched core logic.

fly".

A second source of variation with respect to the way arguments are constructed, concerns the degree to which an argument's internal structure is explicit in the representation of the argument. The situation here is roughly analogous to the way inferences are represented in, say, **CL**. At the level of logical consequence, the inference from the premises $\neg q \wedge \neg p$ and $\neg p \supset r$ to the conclusion r is represented as $\{\neg q \wedge \neg p, \neg p \supset r\} \vdash r$. But in a natural deduction of r given these premises, we can represent this inference in a more 'structured' way as a proof in which we apply rules for implication and conjunction:

1.	$\neg q \wedge \neg p$	Premise
2.	$\neg p \supset r$	Premise
3.	$\neg p$	1; $\wedge$-Elimination
4.	r	2,3; $\supset$-Elimination

The latter representation is more informative as to the internal structure of the inference. Of course, there are plenty of other possible representations of the internal structure of this inference, each of which highlights different aspects of the structure of an inference (e.g. in a sequent calculus or tableau system). The same holds true for the ways in which arguments are constructed relative to a knowledge base in a structured argumentation framework. The systems mentioned above behave as listed in Table 2 with respect to the way they represent the internal structure of arguments.[2] Let us illustrate this with the ASPIC$^+$ system.

Example 1 (Argument construction in ASPIC$^+$). *Lower case letters $a, b, \ldots$ are logical atoms in our formal language. Capital letters $A_1, A_2, \ldots$ represent arguments. Suppose our knowledge base contains the strict premise a, and the defeasible rules $a \Rightarrow b$, $b \Rightarrow c$, and $a \Rightarrow \neg c$. The following arguments can then be constructed:*

$A_1:$	$\langle a \rangle$	$A_3:$	$A_2 \Rightarrow c$
$A_2:$	$A_1 \Rightarrow b$	$A_4:$	$A_1 \Rightarrow \neg c$

---

[2]Most structured accounts of argumentation (such as ASPIC$^+$, DA, DeLP and SBA) operate on the level of arguments (e.g. derivations, a set of premises, etc) which support a specific conclusion. In contrast, ABA operates at a higher level of abstraction, since attacks are defined directly on the level of sets of assumptions instead of on the level of individual arguments for specific conclusions. There are some formulations of ABA that define attacks on the level of individual arguments (e.g. [65]). However, since attacks are only possible *on* assumptions, these formulations are equivalent (cf. also [147]). ABA can thus be viewed as operating on the level of equivalence classes consisting of arguments generated using the same assumptions. Other systems make the conclusion explicit in the argument but still abstract from (some aspects of) the actual derivation (DeLP, DA, SBA). Only DL and ASPIC systems make the argument structure fully explicit.

System	Argument structure
ABA	Arguments are sets Γ of (defeasible) assumptions
ASPIC$^+$	Arguments A are proof trees built on the basis of premisses ($\langle\gamma\rangle$) and strict ($A_1,\ldots,A_n \to \phi$) or defeasible ($A_1,\ldots,A_n \Rightarrow \phi$) inference rules
DA	Arguments are pairs (Γ,ϕ) of premise sets Γ and conclusions ϕ, where Γ is consistent and $\subseteq$-minimal with respect to entailment in the deductive core logic ($\vdash$), and where $\Gamma \vdash \phi$
DeLP	Arguments are pairs (Γ,φ) of a set of defeasible rules Γ and a conclusion φ derivable from Γ and the strict information
DL	Similar to ASPIC$^+$, except that it represents undercutting defeaters with arrows ($A \rightsquigarrow \phi$)
SBA	Arguments are Gentzen-style sequents $\Gamma \Rightarrow \phi$ derivable in an underlying sequent calculus of some deductive core logic.

Table 2

A_1 simply states the piece of information represented by a. A_2 concludes that b based on A_1 and the rule $a \Rightarrow b$. A_3 extends A_2 to reach the conclusion c on the basis of the additional rule $b \Rightarrow c$ (which can be applied to the conclusion b of A_2). Likewise, A_4 concludes $\neg c$ based on the premise A and the rule $a \Rightarrow \neg c$.

In the examples below, we will usually follow the ASPIC$^+$ representation of arguments. This is because the framework is well-studied and very expressive (witness Table 1), and because its notation and argument representation is transparent and perspicuous (witness Table 2 and Example 1).

In our brief overview of accounts of argument construction in structured argumentation, we did not yet mention where and how the notion of argument strength comes into play. We now set out to do exactly that.

1.2 Premisses as a source of argument strength

Many of the systems considered allow for a stratification of the premisses in terms of their strength. The idea can already be found in one of the first formal systems of non-monotonic reasoning: reasoning with maximally consistent subsets of a stratified set Prem of premisses [39, 23]. The question of how to measure and represent the strength of a premise depends heavily on the application context. We give some examples.

Quantitative rankings. We start with examples of direct numerical rankings π :

Prem $\rightarrow$ N where N is a set of numbers (such as $\mathbb{N}, (0,1), [0,1]$, etc.). In the context of epistemic reasoning one may want to numerically rank premises, e.g. according to how likely or probable they are [126, 22], how trustworthy their sources are [121], or according to the degree of reliability of the method by means of which the information has been obtained or the authority which issued the information [18]. In the context of social choice one may numerically represent how many people voted for a specific option [98].

Qualitative rankings. Sometimes it may be preferable to work with qualitative comparisons. One may, for instance, impose an order over the given premises $\preceq \subseteq$ Prem $\times$ Prem. Related ideas are prominent in non-monotonic logic [99] as well as in epistemology [80, 140] where the order reflects the comparative plausibility of a given piece of information.

Since arguments are often based on multiple (defeasible) premises, in order to compare the strength of arguments based on their premises, one needs to lift the quantitative or qualitative comparisons above from premises to sets of premises. Typical candidates are: lexicographic liftings [42, 21], the weakest link or Smyth ordering [24], the Plotkin [24] or max-min ordering, or the strongest link or Hoare ordering [24].

Example 2 (Liftings of orderings on the premises). *Let $x_i \prec y_i$ (resp. $x_i \preceq y_i$) denote that x_i is strictly less preferred (resp. less or equally preferred) than y_j. Consider the set* DefPrem $= \{p_1, q_1, s_1, \ldots, p_5, q_5, s_5\}$ *of defeasible premises, for which $x_i \prec y_j$ [$x_i \preceq y_i$] iff $i < j$ [$i \leq j$] where $x, y \in \{p, q, s\}$. That is, we have the following stratified sets:*

$$p_1, q_1, s_1 \quad \prec \quad p_2, q_2, s_2 \quad \prec \quad p_3, q_3, s_3 \quad \prec \quad p_4, q_4, s_4 \quad \prec \quad p_5, q_5, s_5$$

Consider the arguments $A_1, \ldots, A_5$ based on the premises listed on the right:

Arguments	Defeasible premises
A_1	p_1
A_2	p_2, p_3
A_3	p_2, p_4, p_5
A_4	p_4, p_5
A_5	p_2, q_4, s_4, p_5

Below we define a number of candidate liftings on $\prec$, where $X, Y \subseteq$ DefPrem and strata are denoted as $S_i = \{p_i, q_i, s_i\}$ for each $1 \leq i \leq 5$. Figure 1 illustrates the applications of these lifting principles to the arguments $A_1 - A_5$.

max-max: $X \prec Y$ *iff for every $x \in X$ there is a $y \in Y$ such that $x \prec y$*

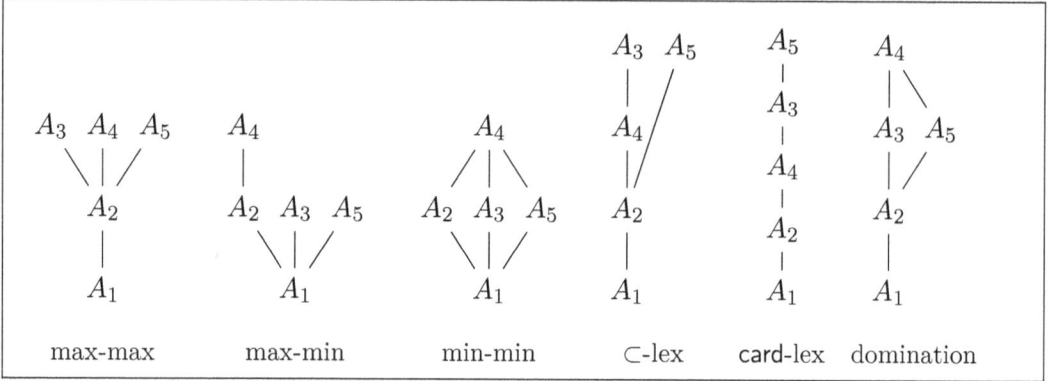

Figure 1: Different lifting principles for Example 2

max-min: $X \prec Y$ iff $x \prec y$ for all $x \in \max(X)$ and all $y \in \min(Y)$

min-min: $X \prec Y$ iff for every $y \in Y$ there is a $x \in X$ such that $x \prec y$

⊂-lex: $X \prec Y$ iff there is $i \in \{1, \ldots, 5\}$ for which $X \cap S_j = Y \cap S_j$ whenever $j > i$ and such that $X \cap S_i \subset Y \cap S_i$

card-lex: $X \prec Y$ iff there is $i \in \{1, \ldots, 5\}$ for which $\mathsf{card}(X \cap S_i) \subset \mathsf{card}(X \cap S_i)$ and $\mathsf{card}(X \cap \bigcup_{j>i} S_j) = \mathsf{card}(Y \cap \bigcup_{j>i} S_j)$

domination: $X \prec Y$ iff for all $x \in X$ there is $y \in Y$ for which $x \preceq y$ and for all $y \in Y$ there is $x \in X$ for which $x \preceq y$.

Probability. There are two basic proposals as to what it means to say that arguments are probable to a certain degree. In the *constellations approach* [101, 89] or *external view* [71] it reflects the probability that the argument is justified in playing a role in a given discursive situation. This approach models an uncertainty about the structural properties of the discursive situation concerning what arguments and counter-arguments should be considered. We will describe it with more detail in Section 3 when discussing argumentative evaluations. In the *epistemic approach* [144, 90, 91, 122] or *internal view* [71] there is no uncertainty about the discursive situation in the sense that we know for sure which arguments and counter-arguments belong to the framework. Here, the probability of an argument is (subjectively) interpreted as the credence an agent has in the argument being acceptable. The main research question concerns the rationality of a probability assignment to arguments in view of the given discursive situation. We will also come back to this when talking about evaluations in Section 3.

Suppose now that an assignment of probabilities to defeasible premisses is given. Where arguments are generated from defeasible premisses and strict rules, one may take the probability of the conjunction of the premisses of an argument to be the probability of the argument in question. For instance, in [90], a probability distribution over the models of a given language is used to assign probabilities to formulas and in turn to arguments based on specific premisses.

Example 3 (Probabilities on assumptions). *Suppose we have the language of (propositional) classical logic with two atoms p and q. We have four classical models in terms of truth assignments to p and q to which we assign probabilities as follows:*

model	$v(p)$	$v(q)$	$P(M)$
M_1	1	1	0.4
M_2	1	0	0.3
M_3	0	1	0.2
M_4	0	0	0.1

The probability of a formula is simply the sum of the probability of all the models in which it is true. For instance, $P(p) = P(M_1) + P(M_2) = 0.7$ and $P(q) = P(M_1) + P(M_3) = 0.6$.

Suppose now we have a defeasible theory with the defeasible premisses p, q and $\neg(p \wedge q)$. We have, among others, the following arguments in (classical) logic-based argumentation: $A_1 = (\{p\}, p)$ and $A_2 = (\{q, \neg(p \wedge q)\}, \neg p)$. The probabilities of A_1 and A_2 are then calculated as the probability of the conjunction of their premisses: $P(A_1) = P(p) = 0.7$ and $P(A_2) = P(q \wedge \neg(p \wedge q)) = P(M_3) = 0.2$.

In the dialogical setting of [136] an account of an argument's chance of construction in the setting of legal reasoning is given in terms of the probability of the premisses it is based on.

Another source of the strength of an argument comes with the quantity and quality of the available evidence that supports it. If I don't know anything about the possible bias of a given coin that looks evenly shaped on both sides, I may form the argument that the coin has a 0.5 chance to land heads based on its appearance. If I additionally learn from a source that she observed that out of the past 1000 throws 500 landed head, my argument can be improved by calling upon the additional evidence and so has more *weight*, since now the available evidence provides more relevant information for the statement. Note that in this case my credence in the coin falling heads and the underlying conditional probability remains unaltered. In the paper "Imprecise Probability and the Measurement of Keynes's 'Weight of Arguments' " (this volume), William Peden proposes a quantitative account of

Keynes' notion of argument weight [97] in terms of Kyburg's system of Evidential Probabilities [100]. Unlike Keynes' notion, Peden allows also for cases where additional relevant evidence may decrease an argument's weight (suppose our source above additionally states that she was drunk while observing the 1000 coin flips). He demonstrates how his account can be used to analyse philosophical problems such as Popper's paradox of ideal evidence [128].

Some authors [29, 124] have argued against the use of probability theory in specific applications of defeasible reasoning. For example, [124] argued that since the probability of a conjunction $A \wedge B$ could be lower than both the probability of A and of B, probability theory precludes inferences from multiple premises that would be logically valid [151]. [29] argue that various assumptions necessary for probabilistic reasoning are not feasible when dealing with uncertainties occurring in deontic reasoning or in the reasoning to an interpretation.

Specificity. Another principle to compare the strength of two arguments based on their premises is to check their degree of specificity [104, 127, 141]. If I know that "Tweety is a bird" I may use the default that "birds fly" to compose an argument that "Tweety flies". However, if I additionally know that "Tweety is a penguin" then I may use the default rule that "penguins do not fly" to infer that "Tweety does not fly". The second argument should be deemed stronger since the more specific information encodes an exceptional situation for the first default. Things get, of course, more complicated if the more specific argument is based on less reliable information. One may wonder whether specificity is best considered as a source of argument strength or rather as giving rise to a specific type of undercut attack (see below).

Tracking specificity can be difficult when considering arguments that have a more complex structure. We give some examples to illustrate this:

Example 4 (Specificity). *We use the ASPIC$^+$ framework to describe several arguments, with square brackets representing a set of arguments that are used to construct a new super-argument using a defeasible rule. (For example, in Case 1, $\langle p \rangle \Rightarrow t$ and $\langle q \rangle \Rightarrow v$ are used to construct an argument with conclusion s using the defeasible rule $t, v \Rightarrow s$.)*

$$1. \; A = \begin{bmatrix} \langle p \rangle \Rightarrow t \\ \langle q \rangle \Rightarrow v \end{bmatrix} \Rightarrow s \; and \; B = \begin{bmatrix} \langle p \rangle \Rightarrow u \\ \langle q \rangle \Rightarrow w \\ \langle p' \rangle \Rightarrow u' \end{bmatrix} \Rightarrow \neg s$$

2. $A = \langle r \rangle \Rightarrow g$ and $B = \langle r \rangle, \langle l \rangle \Rightarrow g \vee s$

3. $A = \langle s \rangle \Rightarrow y \rightarrow a \Rightarrow m$ and $B = \langle s \rangle \Rightarrow \neg m$.

In Case 1, B is more specific than A if we only take into account the premises used in the construction of these arguments: B relies on p, q, and p', while A relies only on p and q. However, the conclusions of A and B were reached via different intermediate steps: t and v in the case of A as opposed to u, w, and u' in the case of B. Given these additional steps in the construction of A and B, are we still allowed to conclude that B is the more specific argument?

In Case 2, the more specific argument B leads to a less specific (logically weaker) conclusion. [87] proposed this case as a problem case for accounts that track specificity (for similar examples also see [35]). Consider two islands, the green island and the sand island. Let $r \Rightarrow g$ represent "ruffed finches live on the green island" while $r, l \Rightarrow g \vee s$ represents "least ruffed finches live on the green or the sand island". If we are dealing with a least ruffed finch, we may want to infer $g \vee s$ on the basis of the more specific argument B, without inferring g on the basis of the less specific argument A.

Case 3 [66, p. 40] indicates that it is not sufficient to consider only the premises of arguments, also sub-conclusions and defeasible links used in the respective arguments are relevant and considered in the literature ([72, 86, 157]). We read $a \Rightarrow m$ as "Adults are usually married", $s \Rightarrow \neg m$ as "Students are usually not married", $s \Rightarrow y$ as "Students are usually young adults", and $y \rightarrow a$ as "Young adults are adults". In this example it is intuitive to consider argument B as more specific than A since from 'being a student' (s) we get 'being a young adult' (y) which is more specific than 'being an adult' (a) on which the last defeasible link in A is based. The example is also instructive if we consider a student that is not a young adult, but an adult. In this case we have reasons to stay agnostic concerning the question whether our student is married. This means that when dealing with normal students we consider the defeasible link $s \Rightarrow \neg m$ as more specific than $a \Rightarrow m$ due to the argument $s \Rightarrow y \rightarrow a$, however, when dealing with exceptional students for whom $\neg y$ holds, this is not the case. Specificity is thus a contextual matter which has to be determined conditional on the specific factual context. This is problematic for many accounts that treat it by ordering defeasible links in a context-independent way and then, on this basis, compare arguments based on this ordering (e.g. [40, 73]).

1.3 Defeasible rules as a source of argument strength

The strength of an argument is influenced not just by the strength of the premisses it is based on, but also by the strength and reliability of the inference rules used in its construction. In view of their truth-preserving (deductive) nature, strict rules are usually considered maximally strong. Things get more interesting once defeasible rules enter the picture. Like defeasible premisses, defeasible rules allow for a stratification in terms of their strength. In analogy to the treatment of defeasible premisses in Section 1.2, we can assign a (qualitative or quantitative) degree of strength to the defeasible rules in our knowledge base. In epistemic reasoning contexts the driving consideration may be one of plausibility, typicality, likelihood, etc. In the context of normative reasoning (such as legal or moral reasoning) defeasible links may represent norms or legal principles which come with degrees of deontic or legal urgency (e.g., in view of the authority which issued them, etc.). As in Section 1.2, lifting principles are necessary when considering arguments that are composed of multiple defeasible rules. This way we gain a comparative notion of argument strength based on the strength of the defeasible links in an argument.

Below we present some lifting principles for defeasible rules suggested in the literature, and we apply these to a 'prioritized' version of Example 1.

Example 5 (Example 1, cont'd). *Suppose $a \Rightarrow b$ has the lowest strength 1, $a \Rightarrow \neg c$ has strength 2 and $b \Rightarrow c$ has the highest strength 3. Let us represent this as:*

$$a \Rightarrow^1 b \qquad a \Rightarrow^2 \neg c \qquad b \Rightarrow^3 c$$

Weakest Link. The weakest link principle says that an argument is as strong as its weakest defeasible elements[3]. Following Examples 1 and 5 where, recall $A_1 = \langle a \rangle$, this means that

$$A_2 : A_1 \Rightarrow^1 b$$
$$\text{and} \qquad \text{are less strong than} \qquad A_4 : A_1 \Rightarrow^2 \neg c$$
$$A_3 : A_2 \Rightarrow^3 c$$

since both A_2 and A_3 make use of the weakest rule $a \Rightarrow^1 b$, whereas the only rule used in A_4 is $a \Rightarrow^2 \neg c$. Weakest link has been argued to be a particular suitable lifting principle for epistemic reasoning [126]. Let us interpret the example as follows:

$(a \Rightarrow^1 b)$: "given all the evidence, he was drunk and driving", according to the report by a not very reliable neighbour

$(b \Rightarrow^3 c)$: "if someone is in the car, he cannot be struck by lightning", according to physics textbooks

$(a \Rightarrow^2 \neg c)$: "given all the evidence, he was struck by lightning", said by the coroner

In this context, it seems reasonable to infer that this person was struck by lightning after all. If one adds the contraposed rule $\neg c \Rightarrow^3 \neg b$ one also obtains that he was

---

[3]When arguments have defeasible rules in common, one can choose to either take into account these common defeasible rules as in e.g. [112] or to ignore shared defeasible rules as in e.g. [44, 160].

not in a car. (Although accepting contraposition in general for defeasible rules is controversial [46].)

Last Link. The last link principle says that an argument is as strong as the last defeasible elements applied in the construction of the argument. In Example 5, as before we get that A_2 is weaker than A_4, but now A_4 is weaker than A_3, since A_3's last rule ($b \Rightarrow^3 c$) is stronger than that of A_4 ($a \Rightarrow^2 \neg c$). As an example from deontic logic (taken from [103]), suppose that a soldier gets the following orders:

($a \Rightarrow^1 b$) : "in winter, turn on the heating", said by a captain

($a \Rightarrow^2 \neg c$) : "in winter, do not open the window", said by a major

($b \Rightarrow^3 c$) : "if the heating is on, open the window", said by a general

In this context, the soldier disobeys the order of the highest ranking order if she reasons according to the weakest link principle. This outcome is avoided if she reasons according to the last link principle (or makes use of contraposition, see [103] for more details).

A complication that arises with the last link lifting concerns cases in which the last link in an argument aggregates several defeasible arguments. In such cases it is not clear what the last defeasible link of the argument is. Take, for instance, the argument

$$A = \begin{bmatrix} \langle p \rangle \Rightarrow u \\ \langle q \rangle \Rightarrow v \\ \langle r \rangle \Rightarrow w \end{bmatrix} \rightarrow u \wedge v \wedge w$$

We can, for instance, opt for the strongest of the last links (called *elitist* approach [112]) or the weakest (called *democratic* approach in [112]).

In systems where strength is represented in the object language, the strength of an argument can be identified with the strength k of its conclusion, say (φ, k), and possibly that of its sub-conclusions —see Section 2 for details.[4]

Probabilities. In Section 1.2 we have discussed probabilistic notions of argument strength. When working with defeasible rules, it makes sense to let the probability

---

[4]For example, [1] incorporates weakest link into DeLP. The weakest link lifting takes the form of generalized *modus ponens*:

$$\frac{(p,k) \quad \dots \quad (p',k') \quad (p \wedge \dots \wedge p' \Rightarrow q, m)}{(q, \min(k, \dots, k', m))}$$

A related approach is [142] which fuzzifies the distinction defeasible/strict of premises and rules with a fuzzy membership function. The degrees of premises and rules in an argument determine the argument's strength, which finally translates into a notion of strength for extensions.

of an argument be determined by both the probability of its premises and the probability of the defeasible rules used in its construction. See [67] for an account based on assumption-based argumentation and [135] for an account based on semi-abstract argumentation frameworks.

As [132] remarks, one has to be careful when considering a probabilistic treatment of argument strength in the presence of defeasible rules: the probability of "most Belgians speak French" is ambiguous between (i) the probability of the truth of the proposition (which is 1, since it is true that most Belgians speak French); and (ii) the probability of the consequent "x speaks French" given the antecedent "x is a Belgian" (which is definitely smaller than 1). Furthermore, the relationship between probabilities of arguments and their dialectical strength is not very clear. Suppose for example that 90 percent of all birds fly and that 80 percent of the chickens on Annie's farm do not fly (where these statements are interpreted in sense (ii) above). Then it seems that we would still like "Miss Prissy is a chicken on Annie's farm and therefore she does not fly" to defeat "Miss Prissy is a bird and therefore she flies", even though the probability of the first argument is lower than the probability of the second argument [132].

These problems are avoided in [146] where a method is described that allows to compute a well-behaved structured argumentation framework in the ASPIC$^+$-family from a given Bayesian network —*well-behaved* in terms of the rationality postulates.[5] The authors study two measures of argument strength that make use of the quantitative information given in the Bayesian network: incremental ones expressing the change in the odds of a hypothesis if the evidence were observed resp. if the relevant premises were considered true, and absolute ones expressing the a posteriori belief in a hypothesis. The Bayesian approach to evidential reasoning is, for instance, very useful in the context of legal reasoning. However, it is often difficult to grasp and to use by non-experts, which is why it is attractive to find more self-explanatory and accessible representations. Other translations into the argumentative setting can be found in [96, 155].[6]

An interesting research question, tackled in a probabilistic setting in [74, 75], concerns the extent to which chaining defeasible rules leads to a decrease of argument strength under an epistemic interpretation.

Directness. Argument strength comparisons based on directness give priority to arguments whose conclusions are "tied closely to the evidence" [104]. For example,

---

[5]The Rationality Postulates are a list of desiderata for structured argumentation systems that these should satisfy in order to be logically well-behaved. They include consistency and logical closure conditions, among others. See [50, 49, 48] for more details, as well as Sec. 3.1.1.

[6]For an approach that goes the other direction and extracts a Bayesian network from a Carneades argumentation framework [77], see [79].

suppose that $a \Rightarrow b$ means "cats are generally aloof to people" and $b \Rightarrow c$ means "aloofness is an indicator of dislike" whereas $a \Rightarrow \neg c$ means "cats generally like people". According to [104], $\langle a \rangle \Rightarrow \neg c$ is to be preferred over $\langle \langle a \rangle \Rightarrow b \rangle \Rightarrow c$ in view of directness considerations.

Summing up our discussion on lifting principles for defeasible rules, it is clear that some such principles, such as directness, are defined purely in terms of the *structure* of arguments. Others, like weakest link, are defined purely in terms of the *strength* of the defeasible rules used in argument construction. Other lifting principles take into account *both* the strength of defeasible rules *and* the general structure of an argument. The last link principle exemplifies such a hybrid approach, as not just the strength of the defeasible rules matters, but also the *order* in which these rules are applied in the construction of the argument.[7]

From the discussion above it is clear that there are many candidate lifting principles, and that these may give rise to quite different outcomes. Most authors agree that the appropriateness of a specific lifting principle mainly depends on the context of application. For instance, Pollock argued that weakest link is appropriate in an epistemic setting [126], while perhaps last link is more appropriate in a legal setting [113].

The lifting principles described above are perhaps best understood as general guidelines rather than full-fledged formal definitions. In the paper "On elitist lifting and consistency in structured argumentation" (this volume) Sjur Dyrkolbotn, Truls Pedersen and Jan Broersen offer a clarifying exposition of the problems that occur when one tries to give a formal account of the weakest link principle. They propose sufficient conditions for specific instantiations of weakest link that avoid these problems, in a quite general (signature-based) approach. In general, it seems that problems (and solutions) related to argument construction often occur because of subtle interactions between the argument construction process, the construction of the defeat relation and the evaluation of arguments. For other examples of such problems see [84, 112].

1.4 Orderings on defeasible information

In the approaches discussed in sections 1.2 and 1.3, (part of) the defeasible information in a knowledge base is assumed to be ordered, where the ordering represents preferences over the different elements in our knowledge base. Picking an appropriate way to order defeasible information is a *representational choice* which may

---

[7]On a related note, in [69] it is argued that arguments of the form $A \Rightarrow B \rightarrow D$ are stronger than $A \rightarrow C \Rightarrow \neg D$ in the context of the defeasible interpretation of $A \Rightarrow B$ as "most As are Bs", and the strict reading of $A \rightarrow C$ as "all As are Cs".

have severe conceptual and technical consequences. Below we list some of the main alternatives.

Linear orders. If not the most popular then certainly the most well-behaved and simplest option is to assume that the knowledge base is mapped into a *total* or *linear order* $\leq$.[8]

Preorders. A pre-order is a binary relation that is reflexive and transitive. It allows for the mutual incomparability of defeasible elements in our knowledge base. In an epistemic setting, this move may be motivated in terms of *epistemic incomparability*: it might be the case that the user who supplies the defeasible knowledge base does not know the actual preferences of each element in the base; enforcing a preference in such cases might lead to unwanted consequences [158]. Another motivation for giving up the totality assumption is *deontic incomparability*. When reasoning with norms or conditional obligations, for instance, different sources of such norms might be incomparable. Suppose we have a Christian soldier who can be given commands by a captain, a general and a priest. Any of the captain's commands is less preferred than those from the general, but both types are incomparable with any command from the priest. A similar reason for incomparability is given by the values with respect to which we compare choices. Suppose for example I want to choose a hotel in Bielefeld. I prefer hotels with a gym and I likewise prefer hotels close to the railway station. If only two hotels exist in Bielefeld, one with a gym but not close to the station, the other close to the railway station but without gym, the two hotels might be considered incomparable in terms of my preferences.

When allowing for incomparable elements in the defeasible knowledge, however, various complications can arise. One class of problems arise within *representational results*. Already in the seminal [63], argumentation theory was motivated as a unifying formalism for defeasible reasoning. This role as unifying formalism depends crucially on its ability to capture many different formalisms for defeasible reasoning, such as default logic, auto-epistemic logic and answer set programming. Various variants of prioritized default logic, however, turned out difficult to represent when considering non-total pre-orders [159, 103]. Another class of problems arises in the context of the *rationality postulates* [48]. In fact most of the research on rationality postulates that has been done for systems of structured argumentation with argument strength is restricted to total orders. Especially crash-resistance and non-interference have not received sufficient attention for non-total orders.

---

[8]Recall that a linear order is antisymmetric, transitive and total.

644

2 The Dialectic Dimension: Attack and Defeat

In Section 1 we looked at the strength of isolated arguments as a function of the support provided by the premisses and inference rules used in their construction. We now turn to the dialectic dimension of argumentation, and study the notion of argument strength in relation to attacks and defeats between arguments.

A typical example of an argument A attacking another argument B is that A concludes (or derives) a formula which is the contrary of a formula in B. By 'contrary', one can mean a pair of an atom p and its negation $\neg p$; or more generally, any pair of formulas ϕ, ψ such that ϕ is classically equivalent to $\neg \psi$. The set of contraries of ϕ is usually denoted $\overline{\phi}$. The set of contraries of a given formula ϕ depends on the application context. The contrariness relation, as expressed by e.g. $q \in \overline{p}$, can be asymmetric so that one can have $p \notin \overline{q}$. This can be used to represent *default negation* (also known as *negation as failure*).

In Section 2.1 we introduce four standard variations of argumentative attack, and we discuss how such attacks may give rise to argumentative defeats in the presence of a concept of argument strength as introduced in Section 1. In Section 2.2 we turn to some further developments aimed at providing a more detailed account of how argument strength comes into play at the dialectic level of argumentation.

2.1 From attack to defeat

Following the bulk of the literature, we start by considering argumentative attack and argumentative defeat as *binary* relations between arguments. Argumentative attack is a purely syntactical relation that expresses the fact that two arguments are in conflict.

Example 6 (Attack and defeat). *Suppose our knowledge base contains a set of strict premisses $\{p \wedge q, u, u'\}$, a set of defeasible premisses $\{p'\}$ and a set of defeasible rules described below. These rules are assigned a name r_i in the formal language, which can be used to define attacks on rules.*

$$r_1 : u \Rightarrow^4 v \qquad r_3 : v \Rightarrow^1 \neg p \qquad r_5 : p' \Rightarrow^3 q'$$
$$r_2 : u' \Rightarrow^3 \neg v \qquad r_4 : p \wedge q \Rightarrow^1 \neg r_2 \qquad r_6 : u \Rightarrow^2 \neg p'$$

As before, superscripts on defeasible rules indicate strength —the higher the number, the stronger the rule. We have, amongst others, the following arguments (written in ASPIC$^+$-style).

$$A = \langle p \wedge q \rangle \to p$$

$$B = \langle u \rangle \Rightarrow^4 v$$

$$C = \langle u' \rangle \Rightarrow^3 \neg v$$

$$D = \langle p' \rangle$$

$$E = B \Rightarrow^1 \neg p$$

$$F = \langle p \wedge q \rangle \Rightarrow^1 \neg r_2$$

$$G = D \Rightarrow^3 q'$$

$$H = \langle u \rangle \Rightarrow^2 \neg p'$$

We list four types of argumentative attack, here presented in the terminology of ASPIC$^+$:

Rebuttal An argument X *rebuts* an argument Y if the conclusion of X is contrary to the conclusion of Y. In the remainder, we take the *contrary* of ϕ to be any formula which is classically equivalent to $\neg\phi$. In Example 6, A and E rebut each other; so do B and C.[9]

Undermining An argument X *undermines* an argument Y if the conclusion of X is contrary to one of the defeasible premises of Y. In Example 6, H undermines G and D.

Undercut An argument X *undercuts* an argument Y if the conclusion of X is the contrary of r_i, where r_i is the last rule applied in the construction of Y. In Example 6, F undercuts C.

Sub-argument attack An argument X *sub-argument attacks* Y (in Y') if it rebuts, undermines, or undercuts a sub-argument Y' of Y. What is counted as a sub-argument depends on the underlying formalism. In ASPIC$^+$ a sub-argument is a sub-tree of an argument's construction tree, e.g. B is a sub-argument of E and of itself. In sequent-based argumentation $\Gamma \Rightarrow \phi$ is a sub-argument of $\Gamma' \Rightarrow \phi'$ if $\Gamma \subseteq \Gamma'$. In Example 6 each of C and D sub-argument attacks E. Similarly, H sub-argument attacks G.

Attacks do not yet take into account the strength of individual arguments. For instance, in Example 6 the arguments B and C attack one another despite the fact that B is clearly stronger. For an attack to be *successful* and *succeed* as a *defeat*, we need to take into account the strength of the arguments involved. Following [110] we take "defeat" to be synonymous to "successful attack".

---

[9] Following Pollock [125, 126], rebuttal is often further restricted so that X only rebuts Y if the conclusion of Y was obtained by applying a defeasible rule. Applied to Example 6, this leads to a disruption of the symmetry of rebutting attack: A still rebuts E, but E no longer rebuts A. This restricted approach to rebuttal has certain advantages, but a discussion of these is beyond the scope of this paper. For more details, see [51, 84].

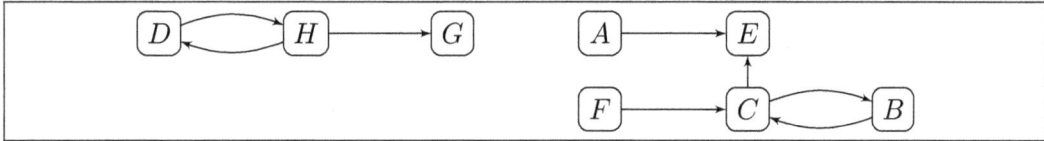

Figure 2: An attack diagram for Example 6.

In Section 1 we listed various ways for determining the strength of an argument in isolation, based on the support provided by its premises and by the rules used in its construction. Given a clear account of how to calculate an argument's individual strength, there are various ways of bringing argument strength into play at the level of argumentative defeat.

First, consider rebuttal. On the account in [112], a rebutting attack by argument A on argument B leads to defeat if A is not strictly weaker than B. [143] demands that A must be (strictly) stronger (in their terminology, *more important*) than B in order for the attack to be successful. Suppose that we adopt the former notion, and that in addition we use the weakest link lifting for determining the strength of the arguments in Example 6. Then, for instance, A defeats E. In ASPIC$^+$ strict arguments (those without defeasible rules) are considered indisputable and so E does not attack back. Moreover, B defeats C since B is stronger than C. Although C rebuts B, this attack is not successful since C is strictly weaker than B.

In general, the considerations about rebuttal-defeats also apply to undermining-defeats. For example, in [112], an undermining attack by argument A on argument B leads to defeat if A is not strictly weaker than B. We note here that in ASPIC$^+$ the orderings for defeasible premises and defeasible rules may be independent from each other. As such, these defeasible premises and defeasible rules may play different roles in determining the strength of arguments. In [112, Def. 21 (Weakest Link)], for example, an argument A without defeasible rules but with defeasible premises (called strict and plausible) is incomparable with an argument B with defeasible rules but without defeasible premises (called defeasible and firm). In view of this, in Example 6, the defeasible and firm H defeats the strict and plausible D, and, vice versa, D defeats H.

Next, consider sub-argument attack. Here the convention is to compare the strength of the attacker X to the strength of the attacked sub-argument Y'. For instance, in Example 6 C sub-argument attacks E in B, and the relevant comparison in terms of strength is between C and B. Since B is stronger than C, we obtain that C does *not* defeat E (in B) even though C is stronger than E (still under the assumption that we use the weakest link lifting).[10] The same goes for the attack

---

[10]With the last link lifting, sub-argument attacks sometimes lead to strange scenarios, since the

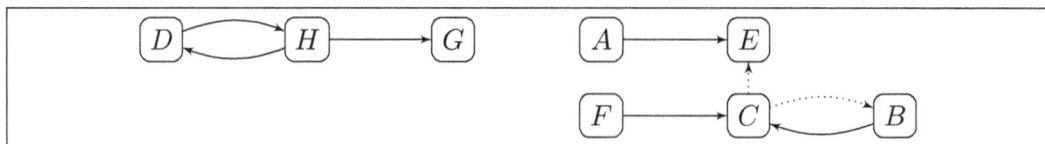

Figure 3: A defeat diagram for Example 6 using preference-Independent undercuts. Attacks that do not result in defeats are represented as dotted arrows.

H to G.

Finally, we turn to undercutting attacks. There is some debate as to what it means for an undercutter to be sufficiently strong for the attack to succeed. Are undercutting attacks successful independently of the strength of the arguments involved? Or should an undercutting attack by A on B lead to defeat only in case A is not strictly weaker (or even strictly stronger) than B? [112] work with a *preference-independent* notion of undercutting defeat according to which undercutting attacks always give rise to defeat. Their motivation, however, has been questioned by [13]. A preference-dependent notion of undercutting attack was recently proposed by [18] in the context of deontic logic, with the aim of modelling a cautious, 'austere' style of reasoning.

A further possible application of the concept of undercutting attacks relates back to the specificity principle as discussed in Section 1, There, we pointed out that comparisons of arguments based on specificity are in a sense orthogonal to comparisons based on the strength of the (defeasible) premises on which the respective arguments are based. The same applies to the strength of the defeasible links on which they are based. Another question is what kind of dialectic behaviour occurs in the context of specificity cases. Should these cases be treated as cases of rebutting defeat with the more specific argument being the stronger one? One complication here is that Dung-style complete semantics give rise to reinstatement, i.e., agents are modelled as having commitments to defended arguments. However, it has been argued [88] that in nested specificity cases reinstatement may be counter-intuitive.

Moreover, argumentative defeat is a problematic way of dealing with specificity in cases like Ex. 4 (item 2) where there is no conflict in the conclusions of the two arguments which constitute a case of specificity. A possible solution is to use undercuts in such situations. Alternatively, inspired by Hempel's principle of maximal specificity [82], [35] introduce *cautious defeaters* as arguments "showing that the total evidence only enables a not so specific conclusion" [35, p. 289] to deal with

---

attacker X may be stronger than the sub-argument Y' but weaker than Y. We list sub-argument attack for analytic purposes. In systems of structured argumentation it is usually integrated in the definition of the other attack forms.

cases such as "least ruffed finches" in the above example and to give an appropriate account of reinstatement in which non-maximally specific arguments cannot be reinstated. In [27] it is argued that in similar cases the more specific argument need not even be factually triggered to exclude the more general one from acceptance. Their example is "If there is a match tonight then John will go to the stadium" versus "If there is a match tonight *and* if John has got enough money then John will go to the stadium". If we only know that there is a match tonight we may want to be cautious since we don't know whether John has enough money and only in the latter hypothetical situation we are –via the second conditional– entitled to conclude that he actually goes to the stadium. This type of hypothetical exclusion mechanism is realized in their account by using conditional logic as a base for structured argumentation.

2.2 Further developments

Reverse attacks. When considering asymmetric attacks (e.g. attacks based on asymmetric notions of contrariness) one can run into scenarios where arguments with contrary (sub)conclusions may not defeat each other, as illustrated in Example 7.

Example 7 (Asymmetric contrariness). *Consider the arguments $A : \langle \top \rangle \Rightarrow^1 q$ and $B : \langle \top \rangle \Rightarrow^2 p$ where q is contrary to p but not vice versa. A attacks but does not defeat B, and B neither attacks nor defeats A. At the level of argumentative defeat, the conflict between A and B goes unnoticed.*

Where A attacks B and B is strictly preferred to A, the attack from A to B is usually considered unsuccessful and so the conflict between A and B is not visible at the level of argumentation defeat. On an extension-based account of argumentation with preferences (cfr. Section 3), this may lead to the result that A and B – despite a clear conflict between them – belong to the same extension. To remedy this defect, some approaches such as [8, 59] introduce a notion of *reverse attack*: if A attacks B and B is strictly preferred to A then B reverse attacks A. On the resulting accounts, the reverse attack by B on A, as opposed to the original attack by A on B, *is* successful. As a result, A and B can no longer belong to the same extension.

Extended argumentation frameworks. In some situations, it might be feasible to allow for the possibility that the information about the strength of an argument is itself *subject to revision*[11]. For example, suppose that $A : \top \Rightarrow^\alpha \neg q$ and $B : \top \Rightarrow^\beta q$ stand for "study a shows that drug x is not safe to use" and "study b shows that

---

[11]In the context of default logic such an approach is said to deal with *dynamic* (as opposed to static) preferences [62, 41]

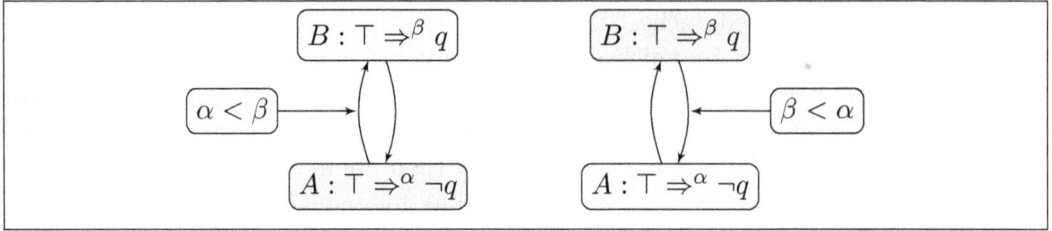

Figure 4: Reasoning with dynamic preferences in extended argumentation frameworks.

drug x is safe to use" respectively. Suppose furthermore that (at moment 1) study a supporting A is deemed of lesser quality (i.e. because it made use of a smaller sample group) than study b supporting B. Then a rational agent might want to conclude that B defeats A but not vice versa since $\alpha < \beta$. Suppose now that the agent later (at moment 2) learns that the scientists conducting study b were actually paid by the pharmaceutical company who developed drug x. In that case, the agent might want to retract $\alpha < \beta$ and possibly replace it by $\beta < \alpha$, resulting in a new defeat graph (A defeats B but not vice versa). This kind of *reasoning about preferences* can be modelled by *extended argumentation frameworks*[12] [110], where arguments are allowed to attack both arguments and other attacks. To see how that allows to model the situation described above, we have to allow for information about preferences to be represented in the object languages. In that case there will be an additional argument $\alpha < \beta$ that attacks the attack relation between A and B, resulting in the graph in Figure 4 (left). We see that $\alpha < \beta$ attacks the attack arrow from A to B. Since $\alpha < \beta$ is unattacked, the attack from A to B is unaccepted and consequently does not have to be taken into account to determine the acceptability of the arguments A and B, resulting in B being accepted. Once $\alpha < \beta$ is replaced by $\beta < \alpha$ in Figure 4 (right), we get exactly the opposite situation, with the attack on the attack from B to A being attacked causing A to be acceptable.

Value-based Argumentation. Instead of defining argument strength (and the defeat relation) in terms of strength of premisses and/or defeasible links, in [20] arguments are associated with *values* they promote. Different orderings on the values may represent different *audiences* of the discursive situation. Defeat is then relative to a given audience, i.e. relative to a given preference ordering on values: an argument A defeats B relative to an audience if A attacks B and the value promoted by B is not more important for our audience than the value promoted by A. The approach has

---

[12] Also called *argumentation framework with recursive attacks* [11] or *higher-order argumentation frameworks* [70].

been generalized and applied in various ways: for instance, in [95] arguments may promote multiple values and non-monotonic preference reasoning is used to compute preferences among arguments; in a similar vein [111] integrates value-based argumentation with higher-order argumentation (see the previous paragraph) to model meta-reasoning about values; and in [61] value-based argumentation is combined with machine learning methods.

Attack strength. So far we treated attack and defeat as *binary* notions in the sense that they either hold or fail between any two arguments. A further refinement is to think of attack and/or defeat as a *gradual* notion. As a motivation (taken from [92]), suppose three persons, Anna, Bernard and Catherine disagree over the colour of a sweater. Anna thinks the sweater is red (A), Bernard thinks that the sweater is pink (B) whereas Catherine thinks that the sweater is blue (C). Then we would perhaps want to say that A and B attack each other to a low degree whereas A and C attack each other to a higher degree, since the difference between red and pink is perhaps subtler than that between red and blue. As with argument strength, different representational formats can be used to represent this *attack strength*. The two main alternatives in the literature are to either opt for a pre-order over the attacks [54, 107] or to assign numerical degrees to attacks [57, 68, 92]. Within the latter approach, [92] allows for arguments to attack each other to a *fuzzy degree* and [101] allows for attacks to have a probability assigned to them. [57, 68], on the other hand, assume that each attack is assigned a *weight* that represents the importance of an attack. A central idea in [57, 68] is that a user can specify an *inconsistency budget* which indicates how much inconsistency (i.e. the sum of the importance of attacks between accepted arguments) a user is prepared to tolerate.

Given the two different approaches in which a notion of strength matters for argumentative defeat, strength of arguments and strength of attacks, one question is how these notions are interrelated or whether one is more primary.

Some papers [94, 56] investigate how to obtain a preference relation over the attacks of an argumentation graph from a given preference ordering over the arguments. For example, [94] investigates how to determine weights on the attack relation on the basis of a given preference relation over arguments by assuming that whenever A attacks B, a stronger preference for A over B will result in a stronger attack from A to B. [56] considers several generalizations of this idea, like using a preference relation instead of a (numerical) weighing function.

Qualifying defeat. Not only attacks, but also defeats can be qualified in terms of their nature and strength. For instance, the Defeasible Logic Programming (DeLP) framework [72, 154] combines two notions of defeat which differ in terms of their strength: *proper defeaters* are strictly better (more specific) than the arguments they defeat, while *blocking defeaters* are incomparable (in terms of specificity) with the arguments they defeat. The difference in strength matters e.g. for defense and reinstatement. An argument with a blocking defeater can only be defended by a proper defeater for the latter, and not by a blocking defeater. (It is also worth mentioning that DeLP departs from most of the literature since Dung [63] by enforcing acyclic defeat paths in the form of finite trees.)

Support relations. Dual to the relation of argumentative attack, one can introduce a relation of argumentative *support* between arguments [55]. (Such a dialectical, *inter-argument* notion of support is not to be confused with the individual, *intra-argument* account of argumentative support discussed in Section 1.) Depending on the context of application, a relation of argumentative support between arguments can be interpreted in various ways. For instance, it has been interpreted as a sub-argument relation [106]; as a relation of deductive support [36]; as a relation of inter-argument evidential support [120]; as an explanatory relation between arguments or between arguments and observations in the context of scientific discourse [139]; and as a relation which represents conditions necessary for the acceptance of a claim [118]. As an example of the latter interpretation, note that turning on the light switch (A) may be a sufficient condition for a light bulb to light up (B) (assuming the electricity works etc.). Consequently, the light bulb's lighting up is necessary for the switch to be turned on. In an argumentation graph, this will mean that whenever B is accepted, A will also be accepted. See also [55, 123] for how some of these systems relate to each other. [5] describes the introduction of support in other dimensions of argumentation processes, besides the dialectical tier. Further extensions with strength of supports are described in [108] and, for abstract dialectical frameworks, in [12, Chap. 5]. Higher-order relations are studied in [36]. Of course, in logic-based or structured argumentation systems, a natural reading of inferential support is already provided by the system with the notion of sub-argument. Indeed, [131] studies to which extent abstract argumentation with support is suitable as an abstraction of ASPIC$^+$ structured argumentation.

3   Argument Evaluation

Following Dung [63], the most influential way to evaluate the status of arguments relative to a given discursive context, i.e., relative to a given attack and/or defeat relation, is via an extensional semantics such as the grounded, preferred, or stable semantics. In Section 3.1 we discuss this approach in relation to the topic of argument strength. In Section 3.2 we turn to ranking-based semantics as a more fine-grained approach to argument evaluation.

3.1 Applying Extensional Semantics over Defeat Graphs

Given a set of arguments $\mathcal{A}$ and an attack relation $Attack \subseteq \mathcal{A} \times \mathcal{A}$, the argumentation framework $(\mathcal{A}, Attack)$ can be interpreted as a directed graph with arguments as nodes and attacks as arrows. Taking into account argument strength and the distinction between attack and defeat (Section 2) the defeat relation $Defeat \subseteq Attack$ induces a new argumentation framework $(\mathcal{A}, Defeat)$.

Given an argumentation framework we can assign acceptability statuses to arguments with the help of argumentation semantics. The most prominent of these are extension-based argumentation semantics,[13] which determine criteria for sets of arguments to represent a rational position within the given discursive situation. One such criterion is that a set of arguments $\mathcal{B} \subseteq \mathcal{A}$ should not contain conflicting arguments. Another is that $\mathcal{B}$ should defend itself against its defeaters — where a set $\mathcal{B}$ *defends* a set $\mathcal{B}'$ if for each defeater of a member of $\mathcal{B}'$, there is a member of $\mathcal{B}$ that defeats it. A set of arguments is a *complete extension* if it is conflict-free and it contains every argument it defends; it is a *preferred extension* (resp. *grounded extension*) if it is $\subset$-maximally (resp. $\subset$-minimally) complete in $\mathcal{A}$; it is a *stable extension* if it is conflict-free and every argument not contained in it is attacked/defeated by one of its members. Whether an argument is *justified*, is then determined in view of the chosen semantics. Some semantics, such as the grounded semantics, determine a unique extension. In this case an argument is justified if it belongs to the unique extension. When dealing with multiple extensions one may consider an argument justified if it belongs to all extensions (skeptical approach) or if it belongs to some extension (credulous approach).[14] Justified conclusions are then defined as conclusions of credulously justified arguments or as conclusions of skeptically justified arguments.[15]

Example 8 (Example 6, cont'd.)**.** *When applying preferred semantics to the defeat diagram in Figure 3 we see that there are two preferred extensions:* $\{D, G, A, F, B\}$ *and* $\{H, A, F, B\}$. *This means that e.g. H is credulously but not skeptically justified*

---

[13]For an introduction to extensional argumentation semantics, see e.g. [10].

[14]Labelling semantics (see [45]) give a more fine-grained approach by offering argument labels such as *in, out* and *undecided* and by specifying relationships in which labelled arguments are supposed to stand: e.g., every argument attacked by an argument that is *in* is *out*, every *out* argument is attacked/defeated by at least one argument that is labelled *in*, if an argument is neither in nor out it is *undecided*, etc.

[15]For the latter there are also different options: ϕ is justified if (a) there is skeptically justified argument with the conclusion ϕ, or (b) in every extension there is an argument with conclusion ϕ. Clearly, if ϕ is justified according to (a) then it is also also justified according to (b) but for many settings this need not hold vice versa. Based on the labelling approach [15] distinguish two principled and incomparable ways to obtain conclusion justifications.

whereas F is skeptically (and thus also credulously) justified. The grounded extension is $\{A, B, F\}$.

3.1.1 Rationality Postulates and Argument Strength

A possible problem with the approach described in Section 3.1 is that information may be lost in the transition from argumentation frameworks based on attacks to argumentation frameworks based on defeat, which may in turn lead to violations of certain rationality postulates for logic-based argumentation.[16] By way of illustration, recall Example 7 from Section 2.2, in which the information that A attacks B is lost in the defeat graph for this example (since A attacks, but does not defeat B). If acceptability is based on defeat, and if a conflict between two arguments amounts to one of these defeating the other, then for Example 7 we obtain the counter-intuitive outcome that A and B belong to the same conflict-free extension, despite the fact that A attacks B. This violates the *consistency* postulate for logic-based argumentation, according to which arguments with contrary conclusions ought not belong to the same extension.[17] As we saw in Section 2.2, consistency can be restored for this example via the introduction of reverse defeat.

Rationality postulates for logic-based argumentation, as introduced by Caminada & Amgoud in [49], are desiderata for logically well-behaved systems of structured argumentation. Besides consistency, Caminada & Amgoud also propose the *closure postulate*, according to which (in ASPIC$^+$ notation) an argument $A_1, \ldots, A_n \rightarrow \varphi$ ought to be justified whenever each of $A_1, \ldots, A_n$ is justified. The motivation for the closure postulate stems from the interpretation of strict rules as truth-preserving or deductive inference tickets: if each of the premises of such a rule is warranted, then so is its conclusion.

Even in the absence of preferences on defeasible premises or defeasible rules, many systems of logic-based argumentation do not satisfy all postulates proposed in the literature. For instance, closure and consistency may fail if rebuttal is unrestricted [53, 51, 84], or if strict rules are not closed under *transposition* [113].[18] In the presence of preferences on defeasible rules, it was shown in [51] that, for the weakest link and last link principles, structured argumentation frameworks using the attack form of unrestricted rebuttal satisfy the postulates of closure and consis-

---

[16] [6] gives several additional problematic examples for the approach using direct defeats and extensional semantics.

[17] The consistency postulate was proposed in [49]. Our formulation is slightly more general and follows [114].

[18] The transposition rule licenses e.g. the inference from the rule $p, q \rightarrow r$ to both the rule $p, \neg r \rightarrow \neg q$ and the rule $\neg r, q \rightarrow \neg p$. The closure and consistency postulates were also proven to hold for knowledge bases that satisfy the similar requirement of contraposition [112, 113].

tency when strict rules are closed under transposition. More generally, it was shown in [112] that ASPIC$^+$ satisfies closure and consistency for frameworks closed under transposition (along with several other side constraints) for any argument ordering that is *reasonable*. A reasonable argument ordering is a set of properties "that one might expect to hold of orderings over arguments composed from fallible and infallible elements" [112, p. 371]. More precisely, an ordering over a set of structured arguments is reasonable iff:

1. strict arguments are maximally preferred, and

2. extending an argument with strict rules or strict premisses does not change the strength of the extended argument.

The generality of this result for reasonable orderings allows the user of the ASPIC$^+$ framework to choose between a variety of lifting principles without having to worry about violations of closure or consistency. Both the last link and the weakest link principle, for instance, result in a reasonable ordering on arguments. But although the conditions proposed in [112] suffice to ensure closure and consistency, other postulates may still fail. For instance, the *non-interference* postulate proposed in [50] may fail for reasonable argument orderings that meet these conditions. The non-interference postulate demands that syntactically disjoint knowledge bases do not *contaminate* one another in the sense that uniting these knowledge bases ought not lead to the unacceptability of arguments that are acceptable in one of the original knowledge bases.[19] In fact the only system in the ASPIC-family for which all rationality postulates were proven to hold when taking into account preferences is ASPIC$^\ominus$ [84], which so far has only been studied under total orders and the weakest link principle.

3.1.2 Preference-Based Extensional Semantics

Motivated by problems presented in Section 3.1.1, the authors of [6, 7] propose to study argument strength on the basis of attack graphs induced by the framework $(\mathcal{A}, Attack)$ instead of defeat graphs induced by the framework $(\mathcal{A}, Defeat)$. Given an argumentation framework $(\mathcal{A}, Attack)$ and a preference order $\preceq \subseteq \mathcal{A} \times \mathcal{A}$, one can define a *dominance ordering* on the power set of $\mathcal{A}$, abusing notation $\preceq \subseteq \mathcal{P}(\mathcal{A}) \times \mathcal{P}(\mathcal{A})$. An extension-based semantics is then defined on the basis of all $\preceq$-maximal sets. Of course, the dominance ordering $\preceq$ has to fulfil specific properties in order to give intuitive outcomes:

- a conflict-free set should always be strictly $\preceq$-preferable to a conflicting set;

---

[19]Two knowledge bases are *syntactically disjoint* if they have no atomic sub-formulas in common.

- the comparison between two sets should be determined by those arguments that are not shared, i.e. $\mathcal{A}_1 \preceq \mathcal{A}_2$ iff $\mathcal{A}_1 \setminus \mathcal{A}_2 \preceq \mathcal{A}_2 \setminus \mathcal{A}_1$.

An equivalent condition for $\mathcal{A}_2 \preceq \mathcal{A}_1$ is that every argument in $\mathcal{A}_2$ is either defeated by an argument in $\mathcal{A}_1$ or reverse attacked. This last requirement seems in a close relation to the reverse-attack approach, and indeed both approaches yield the same results [7, Thm. 5]. Similar ideas where explored in [44] and in the context of assumption-based argumentation in [156]. Furthermore, preference-based extensional semantics are used in various applications in [58, 161].

3.1.3 Uncertainty and Argumentative Evaluations

We recall that the constellations approach (see Section 1.2) models uncertainty concerning structural properties of the discursive situation expressed in the argumentation framework. In [101] the authors work with probability distributions both on the set of arguments and on the attacks between them. In [90, 135] only uncertainty in the question whether arguments are (justified to be) part of the discursive situation is considered. One can now use the given probabilities to calculate how likely it is that an argument is accepted relative to a given semantics. We demonstrate the procedure in the framework of deductive argumentation (cfr. Section 1.1).

Example 9 (Constellations approach). *Suppose we have a defeasible knowledge base with two premisses p and $\neg p$. Suppose further that we consider p to have the probability 0.7 and $\neg p$ the probability 0.3. In (classical) deductive argumentation we have, inter alia, the following two arguments $A_1 = (\{p\}, p)$ and $A_2 = (\{\neg p\}, \neg p)$ standing in a rebuttal-relation to each other. We consider the argumentation framework $\mathcal{G} = (\mathcal{A}, Attack)$ with the set of arguments $\mathcal{A} = \{A_1, A_2\}$ and mutual attacks between A_1 and A_2.*

Argument A_1 has the probability 0.7 while that of A_2 is 0.3. The interpretation according to the constellations approach is that this probability signifies the odds with which the respective argument is part of an arbitrary full sub-graph of the given argumentation framework. The full sub-graphs of our argumentation framework $\mathcal{G}$ are $(\mathcal{G} =) \mathcal{G}_1 = (\{A_1, A_2\}, \{(A_1, A_2), (A_2, A_1)\})$, $\mathcal{G}_2 = (\{A_1\}, \emptyset)$, $\mathcal{G}_3 = (\{A_2\}, \emptyset)$, and $\mathcal{G}_4 = (\emptyset, \emptyset)$.[20] The probability of a sub-graph $\mathcal{H} = (\mathcal{A}_\mathcal{H}, Attack_\mathcal{H})$ is a function of the probability of arguments:

$$P(\mathcal{H}) = \prod_{A \in \mathcal{A}_\mathcal{H}} P(A) \cdot \prod_{A \notin \mathcal{A}_\mathcal{H}} (1 - P(A)).$$

---

[20]Given a graph $\mathcal{G} = (\mathcal{A}, \mathcal{R})$, $\mathcal{H} = (\mathcal{A}', \mathcal{R}')$ is a full sub-graph of $\mathcal{G}$ iff $\mathcal{A}' \subseteq \mathcal{A}$ and $\mathcal{R}' = \mathcal{R} \cap (\mathcal{A}' \times \mathcal{A}')$.

In our case, $P(\mathcal{G}_1) = 0.7 \cdot 0.3 = 0.21$, $P(\mathcal{G}_2) = 0.7 \cdot (1 - 0.3) = 0.49$, $P(\mathcal{G}_3) = (1 - 0.7) \cdot 0.3 = 0.09$, and $P(\mathcal{G}_4) = (1 - 0.7) \cdot (1 - 0.3) = 0.21$.

Next we consider the grounded semantics with the goal to determine each argument's likelihood of being grounded (that is a member of the grounded extension) relative to our sample space, the set $\{\mathcal{G}_1, \mathcal{G}_2, \mathcal{G}_3, \mathcal{G}_4\}$ of full sub-graphs of our initial argumentation framework $\mathcal{G}$. For this we simply sum up the probabilities of those graphs in which the argument in question is grounded, formally:

$$P_{grounded}(A) = \sum \{P(\mathcal{H}) \mid \mathcal{H} \text{ is full sub-graph of } \mathcal{G} \text{ in which } A \text{ is grounded}\}$$

In our case: $P_{grounded}(A_1) = P(\mathcal{G}_2) = 0.49$ and $P_{grounded}(A_2) = P(\mathcal{G}_3) = 0.21$.

We now move our attention to the epistemic approach. Recall that in this approach there is no uncertainty concerning the structure of the discourse or, technically, the structure of the underlying argumentation framework. We start with an argumentation framework $(\mathcal{A}, Attack)$ and a probability function $P : \mathcal{A} \to [0, 1]$ that assigns to each argument the degree to which we assume it to be acceptable. Given the additional dialectic structure underlying the argumentation framework we can then study properties of such probability distributions. Of specific interest is the *epistemic extension* of an argumentation framework in view of P: i.e., the set of all arguments A for which $P(A) > 0.5$. One may think of the epistemic extension as representing all arguments an agent has a doxastic commitment to. For instance, we may require *coherence*: the beliefs of an agent cohere with the given dialectic information in that for any two arguments $A, B \in \mathcal{A}$ we have $P(A) \leq 1 - P(B)$ if A attacks B. A less demanding requirement concerning cases where A attacks B is *rationality*: if $P(A) > 0.5$ then $P(B) \leq 0.5$ (if the agent believes in A then she should not believe in B given the attack). One may also demand that non-attacked arguments should have probability 1 (resp. ≥ 0.5) (this is called *founded* (resp. *semi-founded*) in [91]). *Optimality* demands that $P(A) \geq 1 - \sum_{B \in Attacker(A)} P(B)$ (where *Attacker*(A) is the set of all arguments that attack A) and so gives a lower bound for the belief in A being accepted in terms of the beliefs concerning its attackers being accepted.

Example 10 (Epistemic approach). *We take another look at Example 3. Suppose attack is defined as follows: (Γ, ϕ) attacks (Θ, ψ) iff $\vdash \phi \leftrightarrow \neg\theta$ for some $\theta \in \Theta$.[21] We give an excerpt of the argumentation framework in Figure 5.*

The epistemic extension for our example contains for instance the arguments $(\{p\}, p), (\{q\}, q)$ and $(\{\neg(p \wedge q)\}, \neg(p \wedge q))$. Note that the set of conclusions for

---

[21]In the context of logic- and sequent-based argumentation this attack form is called *Direct Undercut* – unlike the usage of the term in ASPIC$^+$ [112] and by Pollock [126].

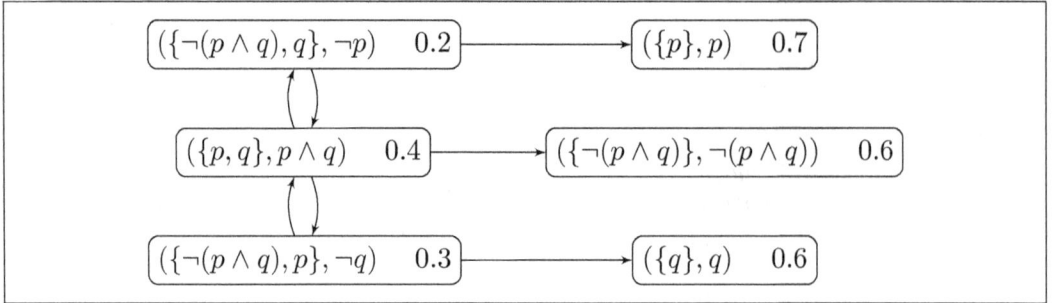

Figure 5: Excerpt of the argumentation framework for Example 10 with probabilities for each argument.

this extension is not consistent. Nevertheless, the underlying probability function of Example 3 satisfies the properties of coherence and (therefore) rationality.

This is not a coincidence, since it can be shown that all consistent probability distributions are coherent [90, Prop. 6], where P is *consistent* iff $\sum_{M \in \mathcal{M}} P(M) = 1$ (for the class $\mathcal{M}$ of classical models over a given signature). Were we to use an inconsistent probability function P' such as (for Example 3) $P'(M_3) = 0.8$ and $P'(M_i) = P(M_i)$ for $i \in \{1, 2, 4\}$, then coherence (and also rationality) would be violated, as witnessed by $P'((\{\neg(p \wedge q), q\}, \neg p)) = P'(M_3) = 0.8 > 1 - P'((\{p\}, p)) = 0.3$. Inconsistent probability functions are studied in more detail in [91].

3.2 Ranking-Based Semantics

Extensional semantics are functions that output a collection of sets of arguments: each set (i.e. each extension) contains arguments that can be reasonably upheld together given an argumentation graph. In a sense, they thus assign a discrete or crisp value to an argument (e.g. a is an element of some extension; a is an element of every extension; a is not an element of any extension)[22]. Ranking-based semantics allow for more fine-grained distinctions in that they impose an ordering on the set of arguments and attacks and/or support relations between them. This ranking of arguments is usually sensitive to (some of) the following factors: the set of an argument's attackers and/or supporters and their ranks, and possibly to a basic strength value with which arguments are equipped. Some rankings in the literature are qualitative (e.g. [2, 145] and the meta-study [38]) while others work with numeric assignments (e.g. [4, 3, 109, 133, 17]).

---

[22]See [14] for more on justification states.

Example 11 (Ranking and defense). *Suppose our set of arguments is $\mathcal{A} = \{A_1, A_2, A_3\}$ where A_1 attacks A_2 and A_2 attacks A_3. In [26] arguments are ranked according to the following categoriser function $cat : \mathcal{A} \to (0, 1]$,*

$$A \mapsto \begin{cases} 1 & \text{if } A \text{ has no attackers} \\ \dfrac{1}{1 + \sum_{B \in Attackers(A)} cat(B)} & \text{else} \end{cases}$$

We have: $cat(A_1) = 1 > cat(A_3) = \frac{2}{3} > cat(A_2) = \frac{1}{2}$.

A similar approach is the burden-based semantics in [2]. Where $i \geq 0$ let $bur_i : \mathcal{A} \to [1, 2]$,

$$A \mapsto \begin{cases} 1 & \text{if } i = 0 \\ 1 + \sum_{B \in Attackers(A)} \frac{1}{bur_{i-1}(B)} & \text{else} \end{cases}$$

Here we get the next sequences of burden numbers (written $\langle bur_0(\cdot), bur_1(\cdot), \ldots \rangle$): $\langle 1, 1, 1, \ldots \rangle$ for A_1, $\langle 1, 2, 2, \ldots \rangle$ for A_2, and $\langle 1, 2, 1.5, \ldots \rangle$ for A_3. By lexicographically comparing these sequences we get the ranking $A_1 \succ A_3 \succ A_2$.

The example illustrates that, unlike in extensional semantics, an argument's strength can be diminished even if it is fully defended (like A_3 in the example above). The next example shows that also the number of attackers matters for our two examples of rankings:

Example 12 (Ranking and number of attacks). *Let our set of arguments be $\mathcal{A} = \{A_1, A_2, B_1, C_1, C_2\}$ where A_1 is attacked by B_1 and A_2 is attacked by both C_1 and C_2. We have*

- *$cat(B_1) = cat(C_1) = cat(C_2) = 1$, $cat(A_1) = \frac{1}{2}$ and $cat(A_2) = \frac{1}{3}$.*

- *the burden number sequences of B_1, C_1 and C_2 are $\langle 1, 1, 1, \ldots \rangle$ while we have $\langle 1, 2, 2, \ldots \rangle$ for A_1 and $\langle 1, 3, 3 \ldots \rangle$ for A_2. By lexicographically comparing the sequences we get the ranking $B_1, C_1, C_2 \succ A_1 \succ A_2$.*

In both cases we see that the argument with more attackers (A_2) is ranked worse than the one with less attackers (A_1).

In the context of ranking-based semantics a great variety of properties have been proposed and studied which help to group the various proposals in the literature. For instance, one expects an argument only to have diminished strength if it is attacked; or, one may require that the strength of an argument is indirectly proportional to the number (and strength) of its attackers (see e.g. [16, 4] for recent investigations into such properties). So far, ranking-based approaches have been mainly studied in the context of abstract argumentation. A notable exception is [162].

4 Argument Accrual

Conclusions generally tend to become more credible the more independent arguments we find in their favour. This, roughly, is the principle of argument accrual. We discuss it in a separate section as it seems to connect to all three tiers of argumentation presented in the introduction.

Most authors tend to agree that arguments accrue,[23] although opinions differ on how to implement argument accrual. We illustrate some approaches by means of an example adopted from [115].

Example 13 (Accrual for defeat). *Max has been invited to a wedding, but since the wedding falls at an inconvenient time, he is inclined not to attend. But now Max finds out that the guests will include his two old aunts, Olive and Petunia, whom he enjoys and who he knows would like to see him. Even though Max would choose not to attend the wedding if only one of the two aunts were going, the chance to see both Olive and Petunia in the same trip offers enough value to compensate for the inconvenience of the trip itself.*

Suppose our knowledge base contains the information that the wedding falls at an inconvenient time (i), that Olive will be present (o), and that Petunia will be present (p). It also contains the defeasible rules $r_1 : i \Rightarrow^3 \neg a$ and $r_2 : o \Rightarrow^2 a$ and $r_3 : p \Rightarrow^2 a$, where a represents Max's attending the wedding.

$$A : \langle i \rangle \Rightarrow^3 \neg a \qquad B : \langle o \rangle \Rightarrow^2 a \qquad C : \langle p \rangle \Rightarrow^2 a$$

In a structured argumentation formalism, a fully automated procedure for argument accrual would, given the arguments B and C, come up with an accrued argument $B \oplus C$ which combines B and C, and which defeats A. In our example, this could be done via a lifting principle which assigns to $B \oplus C$ a degree of priority higher than 3. That would require a lifting principle capable of *raising* the strength of arguments so that accrued arguments are stronger than their constituents. For example, the sum of the strengths of B and C (2 in each case) would assign a strength 4 to their accrual $B \oplus C$. (Clearly, none of the lifting principles we saw in Section 1 will do.) It is a hard and open question whether there are lifting principles adequate for this job, since the mechanism underlying reason accrual is far from clear. As Prakken [129] points out, there are cases in which the accrued argument is in fact *weaker* than its constituents. By way of example he considers two arguments not to go jogging.

---

[23] A notable exception is Pollock: "I doubt that reasons do accrue. If we have two separate undefeated arguments for a conclusion, the degree of justification for the conclusion is simply the maximum of the strengths of the two arguments. This will be my assumption" [126, p. 102].

The first is that it is hot, and the second that it is raining. Each of these arguments may be stronger than the accrued argument for the same conclusion, since we may in fact be more inclined to go jogging if it is hot *and* raining.

A different strategy is to 'manually' add further defeasible rules to our knowledge base for representing accrual. For instance, the rule $r_4 : o, p \Rightarrow^4 a$ allows us to construct the argument $\langle o \rangle, \langle p \rangle \Rightarrow^4 a$, which defeats A as desired. Alternatively, and following Prakken [129], we might treat accrual via undercuts in a setting in which the strength of defeasible rules is not explicitly represented. Omitting superscripts for defeasible rules, we could, for instance, add the rule $r_5 : o, p \Rightarrow \neg r_1$, so that, as a result, the argument $\langle o \rangle, \langle p \rangle \Rightarrow \neg r_1$ undercuts the argument $\langle i \rangle \Rightarrow \neg a$.

Thomas Gordon suggests a different approach to argument accrual in the paper "Defining argument weighing functions" (this volume). Within the structured argumentation approach for 'balancing' arguments from [78], Gordon investigates a number of functions assigning weights to arguments relative to structural properties of argumentation frameworks, and relative to the acceptability status of other arguments in these frameworks. Simple cases of accrual are represented via weighing functions on *cumulative arguments* —arguments the strength of which increases with the number of acceptable premises. Non-cumulative cases of accrual, such as Prakken's jogging example, can also be dealt with via the more intricate notion of a *factorized weighing function*, which measures an argument's weight in proportion to a number of relevant factors, such as "it is hot" and "it is raining" in the jogging example. As Gordon admits, it is currently unclear which rationality constraints apply to weighing functions of the kind he proposes. What *is* clear, is that his account of 'balancing' pros and cons via weighing functions offers interesting prospects for the study of argument accrual.

Within the setting of DeLP, argument accrual has been studied via the concepts of partial and combined attacks on accrued structures which combine arguments with identical conclusions [105, 76]. Via combined attacks two or more arguments may jointly defeat an accrued structure even if none of these arguments on their own can defeat the structure in question. Similarly, Verheij tackles accrual via the notion of *compound defeat*, according to which collections of arguments may defeat other collections of arguments [150, 149]. In the context of abstract argumentation we can find similar approaches where several arguments may jointly attack arguments (or sets of arguments) [117, 33].

In the context of weighted abstract argumentation frameworks (see Section 2), some notions of an accrual of defending arguments have been defined. We illustrate these with an example.

Example 14 (Accrual for defense). *Suppose we have the argumentation framework*

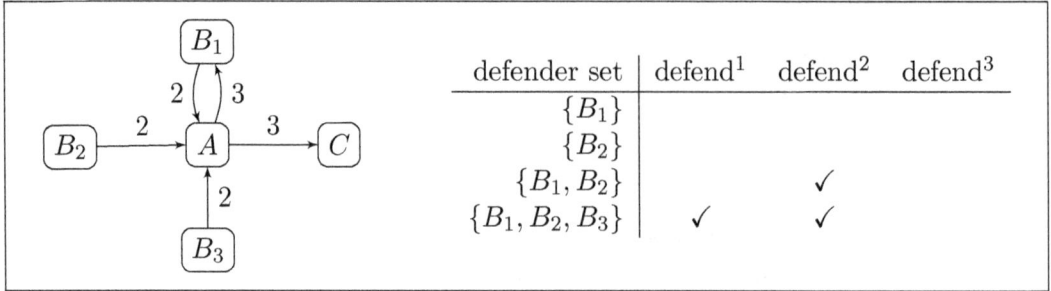

defender set	defend[1]	defend[2]	defend[3]
$\{B_1\}$			
$\{B_2\}$			
$\{B_1, B_2\}$		✓	
$\{B_1, B_2, B_3\}$	✓	✓	

Figure 6: The weighted argumentation framework for Example 14. In the table we indicate whether the set on the left defends C according to the different notions of defense.

with weighted attacks in Figure 6 (left).

Where $W(A, B)$ denotes the weight of the attack from A to B, we can with [31] define the weight of a set of arguments $\mathcal{A}$ attacking a single argument A by $W(\mathcal{A}, A)$ $= \sum_{B \in \mathcal{A}} (B, A)$. Similarly, we can define the weight of an argument's A attack on a set of arguments $\mathcal{A}$ by $W(A, \mathcal{A}) = \sum_{B \in \mathcal{A}} W(A, B)$.[24] We now compare three ways of defining a notion of defense in weighted abstract argumentation, where the first two realize a notion of accrual:

1. According to [31], $\mathcal{A}$ defends[1] B iff for every A that attacks B we have: $W(A, \mathcal{A} \cup \{B\}) \leq W(\mathcal{A}, A)$.

2. According to [57], $\mathcal{A}$ defends[2] B iff for every A that attacks B we have: $W(A, B) \leq W(\mathcal{A}, A)$.

3. According to [107], $\mathcal{A}$ defends[3] B iff for every A that attacks B there is a $C \in \mathcal{A}$ for which $W(A, B) \leq W(C, A)$.

In the table in Figure 6 we give an overview of how these notions apply to our example.

Argument accrual raises a number of difficult open questions, and there is little convergence as to the methods by which to address them. Different approaches highlight interesting aspects of accrual, but at the moment there is no consensus regarding how the strengths of a number of arguments $A_1, \ldots, A_n$ with identical conclusions relate to the strength of their accrual $A_1 \oplus \ldots \oplus A_n$.

---

[24] We suppose in our example that weights are numbers (canonically ordered) and that weights are aggregated by summing them up. The approach in [31] is more general in that it allows to work with arbitrary C-semirings [30].

5 Representational Results

One of the motivations of formal argumentation is to act as a unifying model of formal approaches to defeasible reasoning[25] which offers conceptually motivated explications of many defeasible reasoning forms and has a strong explanatory power (see e.g. [47, 138]). In this capacity, representational results that establish connections between formal (instantiated) argumentation and other formalisms for defeasible reasoning are important to substantiate the claim of formal argumentation as a unifying formalism. For example, in [37] it was shown that assumption-based argumentation can represent auto-epistemic and default logic.

Preferences play an important role in formal models of defeasible reasoning (see e.g. [43, 62]). However, the relation between preferences in defeasible reasoning and argument strength is far from clear. [160] offers representation results for prioritized default logic [40] based on totally ordered preferences within ASPIC$^+$ but shows that several complications have to be overcome to obtain this representation. In [159], further problems and their solutions are discussed for the case where the preference order is not totally ordered. Similar representation results for totally ordered default theories have been presented in [40] for prioritized normative reasoning [81] and the Brewka-Eiter approach to prioritized default logic [42]. For these representations, no generalizations to the case for non-totally ordered default theories have been published yet.

Another important open question concerns the relation between argumentation-based approaches to preferential reasoning on the one hand, and model-theoretic approaches on the other hand. Arguably, the best known representative of the model-theoretic perspective is the 'KLM' semantics [99]. On this semantics, inferences are evaluated by checking whether their conclusion holds true at the best premise-states, where some order on states is given in advance. Some properties of this semantics have been studied from an argumentative perspective in [137]. The KLM semantics also bears close connections to Verheij's *case-based* approach to argumentation [152, 153]. Arguments on Verheij's account are premise-conclusion pairs of the form (φ, ψ). They are evaluated against the backdrop of a case model: a pair $(C, \preceq)$ of a finite set of formulas ('cases') C and an ordering $\preceq$ on C. An argument (φ, ψ) is presumptively valid if ψ is true in some best φ-case, where 'best' means 'maximal with respect to $\preceq$'. On the KLM semantics, inferences are evaluated by checking whether their conclusion holds true at the best premise-states, just as Verheij evaluates arguments by checking whether their conclusion holds true at the best premise-cases. Still, there are important differences between both approaches.

---

[25]This motivation was already present in [63].

For instance, cases on Verheij's account are required to be logically incompatible, and the ordering $\preceq$ is required to be a total pre-order so as to ensure that probabilities can be assigned to cases.

An alternative line of research on the connection between preferential reasoning and formal argumentation is the investigation of various postulates for non-monotonic reasoning as formulated by [99] for argumentation formalisms. Such studies have been carried out for ASPIC$^+$ [102, 64] and ABA [85, 60].

From this short discussion it is clear that there are important open questions concerning the representation of preferential reasoning and argument strength in argumentation-based approaches to defeasible reasoning on the one hand, and some of the paradigmatic approaches in the field of non-monotonic reasoning on the other. Besides the well-known accounts of default logic and preferential semantics, there is a multitude of other approaches to prioritized defeasible reasoning – e.g. [134, 148] – whose representation in formal argumentation has not been investigated yet.

Another open question concerns the comparison of the expressive power of different frameworks of structured argumentation. For instance, while it is known that without priorities, ASPIC$^+$ and ABA are equi-expressive [130, 83], the question is open for prioritized versions.

References

[1] Teresa Alsinet, Carlos Iván Chesñevar, Lluis Godo, and Guillermo Ricardo Simari. A logic programming framework for possibilistic argumentation: Formalization and logical properties. *Fuzzy Sets and Systems*, 159(10):1208–1228, 2008.

[2] Leila Amgoud and Jonathan Ben-Naim. Ranking-based semantics for argumentation frameworks. In *International Conference on Scalable Uncertainty Management*, pages 134–147. Springer, 2013.

[3] Leila Amgoud and Jonathan Ben-Naim. Evaluation of arguments from support relations: Axioms and semantics. In *Proceedings of the International Joint Conference on Artificial Intelligence, IJCAI*, pages pp–900, 2016.

[4] Leila Amgoud, Jonathan Ben-Naim, Dragan Doder, and Srdjan Vesic. Acceptability semantics for weighted argumentation frameworks. In *Proceedings of the International Joint Conference on Artificial Intelligence, IJCAI*, volume 2017, 2017.

[5] Leila Amgoud, Claudette Cayrol, Marie-Christine Lagasquie-Schiex, and P. Livet. On bipolarity in argumentation frameworks. *International Journal of Intelligent Systems*, 23(10):1062–1093, 2008.

[6] Leila Amgoud and Srdjan Vesic. Repairing preference-based argumentation frameworks. In *Proceedings of the International Joint Conference on Artificial Intelligence, IJCAI*, pages 665–670, 2009.

[7] Leila Amgoud and Srdjan Vesic. Generalizing stable semantics by preferences. In *Computational Models of Argument*, pages 39–50, 2010.

[8] Leila Amgoud and Srdjan Vesic. Rich preference-based argumentation frameworks. *International Journal of Approximate Reasoning*, 55(2):585–606, 2014.

[9] Ofer Arieli and Christian Straßer. Sequent-based logical argumentation. *Argument & Computation*, 6(1):73–99, 2015.

[10] Pietro Baroni, Martin Caminada, and Massimiliano Giacomin. An introduction to argumentation semantics. *The Knowledge Engineering Review*, 26(4):365–410, 2011.

[11] Pietro Baroni, Federico Cerutti, Massimiliano Giacomin, and Giovanni Guida. AFRA: Argumentation framework with recursive attacks. *International Journal of Approximate Reasoning*, 52(1):19–37, 2011.

[12] Pietro Baroni, Dov Gabbay, Massimiliano Giacomin, and Leendert van der Torre, editors. *Handbook of Formal Argumentation*. College Publications, 2018.

[13] Pietro Baroni, Massimiliano Giacomin, and Giovanni Guida. On the notion of strength in argumentation: overcoming the epistemic/practical dichotomy. In *Proceedings of the 2001 ECSQARU Workshop: Adventures in Argumentation Toulouse, France*, pages 1–8, 2001.

[14] Pietro Baroni, Massimiliano Giacomin, and Giovanni Guida. Towards a formalization of skepticism in extension-based argumentation semantics. In *Proceedings of the 4th Workshop on Computational Models of Natural Argument (CMNA 2004), Valencia, Spain*, pages 47–52. Citeseer, 2004.

[15] Pietro Baroni, Guido Governatori, and Régis Riveret. On labelling statements in multi-labelling argumentation. In *European Conference on Artificial Intelligence (ECAI)*, pages 489–497, 2016.

[16] Pietro Baroni, Antonio Rago, and Francesca Toni. How many properties do we need for gradual argumentation? In *AAAI Conference on Artificial Intelligence*, 2018.

[17] Pietro Baroni, Marco Romano, Francesca Toni, Marco Aurisicchio, and Giorgio Bertanza. Automatic evaluation of design alternatives with quantitative argumentation. *Argument & Computation*, 6(1):24–49, 2015.

[18] Mathieu Beirlaen, Jesse Heyninck, and Christian Straßer. Structured argumentation with prioritized conditional obligations and permissions. *Journal of Logic and Computation*, 2018.

[19] Trevor Bench-Capon, Henry Prakken, and Giovanni Sartor. Argumentation in legal reasoning. In *Argumentation in artificial intelligence*, pages 363–382. Springer, 2009.

[20] Trevor JM Bench-Capon. Persuasion in practical argument using value-based argumentation frameworks. *Journal of Logic and Computation*, 13(3):429–448, 2003.

[21] Salem Benferhat, Claudette Cayrol, Didier Dubois, Jerome Lang, and Henri Prade. Inconsistency management and prioritized syntax-based entailment. In *Proceedings of the International Joint Conference on Artificial Intelligence, IJCAI*, volume 93, pages 640–645, 1993.

[22] Salem Benferhat, Didier Dubois, and Henri Prade. Argumentative inference in uncer-

tain and inconsistent knowledge bases. In *Uncertainty in Artificial Intelligence, 1993*, pages 411–419. Elsevier, 1993.

[23] Salem Benferhat, Didier Dubois, and Henri Prade. Some syntactic approaches to the handling of inconsistent knowledge bases: A comparative study. Part 2: The prioritized case. In Ewa Orlowska, editor, *Logic at work: Essays Dedicated to the Memory of Helena Rasiowa*, pages 473–511. Springer, 1999.

[24] Salem Benferhat, Sylvain Lagrue, and Odile Papini. Reasoning with partially ordered information in a possibilistic logic framework. *Fuzzy Sets and Systems*, 144(1):25–41, 2004.

[25] P. Besnard, A. Garcia, A. Hunter, S. Modgil, H. Prakken, G. Simari, and F. Toni (eds.). Special issue: Tutorials on structured argumentation. *Argument and Computation*, 5(1), 2014.

[26] P. Besnard and A. Hunter. A logic-based theory of deductive arguments. *Artificial Intelligence*, 128(1):203–235, 2001.

[27] Philippe Besnard, Éric Grégoire, and Badran Raddaoui. A conditional logic-based argumentation framework. In *International Conference on Scalable Uncertainty Management*, pages 44–56. Springer, 2013.

[28] Philippe Besnard and Anthony Hunter. Argumentation based on classical logic. In *Argumentation in Artificial Intelligence*, pages 133–152. Springer, 2009.

[29] Tarek R Besold, Artur d'Avila Garcez, Keith Stenning, Leendert van der Torre, and Michiel van Lambalgen. Reasoning in non-probabilistic uncertainty: Logic programming and neural-symbolic computing as examples. *Minds and Machines*, 27(1):37–77, 2017.

[30] Stefano Bistarelli, Ugo Montanari, and Francesca Rossi. Semiring-based constraint satisfaction and optimization. *Journal of the ACM*, 44(2):201–236, 1997.

[31] Stefano Bistarelli, Fabio Rossi, and Francesco Santini. A novel weighted defence and its relaxation in abstract argumentation. *International Journal of Approximate Reasoning*, 92:66–86, 2018.

[32] J.Anthony Blair. Premise adequacy. In F.H. van Eemeren, R. Grootendorst, J.A. Blair, and C. Willard, editors, *Argumentation: Perspectives and Approaches, Vol. 2*, pages 191–202. Amsterdam: Sic Sat., 1995.

[33] Alexander Bochman. Collective argumentation and disjunctive logic programming. *Journal of logic and computation*, 13(3):405–428, 2003.

[34] Alexander Bochman. Propositional argumentation and causal reasoning. In *International Joint Conference on Artificial Intelligence*, volume 19, page 388, 2005.

[35] Gustavo Adrián Bodanza and Claudio Andrés Alessio. Rethinking specificity in defeasible reasoning and its effect in argument reinstatement. *Information and Computation*, 255:287 – 310, 2017.

[36] Guido Boella, Dov M. Gabbay, Leendert W. N. van der Torre, and Serena Villata. Support in abstract argumentation. In *Computational Models of Argument*, pages 111–122, 2010.

[37] Andrei Bondarenko, Phan Minh Dung, Robert A Kowalski, and Francesca Toni. An abstract, argumentation-theoretic approach to default reasoning. *Artificial intelligence*, 93(1-2):63–101, 1997.

[38] Elise Bonzon, Jérôme Delobelle, Sébastien Konieczny, and Nicolas Maudet. A comparative study of ranking-based semantics for abstract argumentation. In *AAAI Conference on Artificial Intelligence*, pages 914–920, 2016.

[39] Gerhard Brewka. Preferred subtheories: An extended logical framework for default reasoning. In *Proceedings of the International Joint Conference on Artificial Intelligence, IJCAI*, volume 89, pages 1043–1048, 1989.

[40] Gerhard Brewka. Adding priorities and specificity to default logic. In *European Workshop on Logics in Artificial Intelligence*, pages 247–260. Springer, 1994.

[41] Gerhard Brewka. Reasoning about priorities in default logic. In *AAAI Conference on Artificial Intelligence*, volume 1994, pages 940–945, 1994.

[42] Gerhard Brewka and Thomas Eiter. Preferred answer sets for extended logic programs. *Artificial intelligence*, 109(1-2):297–356, 1999.

[43] Gerhard Brewka, Ilkka Niemela, and Miroslaw Truszczynski. Preferences and non-monotonic reasoning. *AI magazine*, 29(4):69, 2008.

[44] Gerhard Brewka, Miroslaw Truszczynski, and Stefan Woltran. Representing preferences among sets. In *AAAI Conference on Artificial Intelligence*, 2010.

[45] Martin Caminada. On the issue of reinstatement in argumentation. In *European Workshop on Logics in Artificial Intelligence*, pages 111–123. Springer, 2006.

[46] Martin Caminada. On the issue of contraposition of defeasible rules. In *Computational Models of Argument*, pages 109–115, 2008.

[47] Martin Caminada. Argumentation semantics as formal discussion. In *Handbook of Formal Argumentation*. College Publications, 2018.

[48] Martin Caminada. Rationality postulates: applying argumentation theory for non-monotonic reasoning. In *Handbook of Formal Argumentation*. College Publications, 2018.

[49] Martin Caminada and Leila Amgoud. On the evaluation of argumentation formalisms. *Artificial Intelligence*, 171:286 – 310, 2007.

[50] Martin Caminada, Walter A Carnielli, and Paul E Dunne. Semi-stable semantics. *Journal of Logic and Computation*, 22(5):1207–1254, 2012.

[51] Martin Caminada, Sanjay Modgil, and Nir Oren. Preferences and unrestricted rebut. *Computational Models of Argument*, 2014.

[52] Martin Caminada, Samy Sá, João Alcântara, and Wolfgang Dvořák. On the equivalence between logic programming semantics and argumentation semantics. *International Journal of Approximate Reasoning*, 58:87–111, 2015.

[53] Martin Caminada and Yining Wu. On the limitations of abstract argumentation. In *Proceedings of the 23rd Benelux Conference on Artificial Intelligence (BNAIC 2011)*, pages 59–66, 2011.

[54] Claudette Cayrol, Caroline Devred, and Marie-Christine Lagasquie-Schiex. Accept-

ability semantics accounting for strength of attacks in argumentation. In *European Conference on Artificial Intelligence (ECAI)*, volume 10, pages 995–996, 2010.

[55] Claudette Cayrol and Marie-Christine Lagasquie-Schiex. Bipolarity in argumentation graphs: Towards a better understanding. *International Journal of Approximate Reasoning*, 54(7):876–899, 2013.

[56] Claudette Cayrol and Marie-Christine Lagasquie-Schiex. From preferences over arguments to preferences over attacks in abstract argumentation: A comparative study. In *Tools with Artificial Intelligence (ICTAI), 2013 IEEE 25th International Conference on*, pages 588–595. IEEE, 2013.

[57] Sylvie Coste-Marquis, Sébastien Konieczny, Pierre Marquis, and Mohand Akli Ouali. Weighted attacks in argumentation frameworks. In *Proceedings of the International Conference on Principles of Knowledge Representation and Reasoning (KR)*, 2012.

[58] Madalina Croitoru, Rallou Thomopoulos, and Srdjan Vesic. Introducing preference-based argumentation to inconsistent ontological knowledge bases. In *International Conference on Principles and Practice of Multi-Agent Systems*, pages 594–602. Springer, 2015.

[59] Kristijonas Čyras and Francesca Toni. ABA+: assumption-based argumentation with preferences. In *Proceedings of the International Conference on Principles of Knowledge Representation and Reasoning (KR)*, pages 553–556, 2016.

[60] Kristijonas Čyras and Francesca Toni. Properties of ABA+ for non-monotonic reasoning. *arXiv preprint arXiv:1603.08714*, 2016.

[61] Artur S D'Avila Garcez, Dov M Gabbay, and Luis C Lamb. Value-based argumentation frameworks as neural-symbolic learning systems. *Journal of Logic and Computation*, 15(6):1041–1058, 2005.

[62] James Delgrande, Torsten Schaub, Hans Tompits, and Kewen Wang. A classification and survey of preference handling approaches in nonmonotonic reasoning. *Computational Intelligence*, 20(2):308–334, 2004.

[63] Phan Minh Dung. On the acceptability of arguments and its fundamental role in non-monotonic reasoning, logic programming and n-person games. *Artificial intelligence*, 77(2):321–357, 1995.

[64] Phan Minh Dung. An axiomatic analysis of structured argumentation with priorities. *Artificial Intelligence*, 231:107–150, 2016.

[65] Phan Minh Dung, Robert A Kowalski, and Francesca Toni. Dialectic proof procedures for assumption-based, admissible argumentation. *Artificial Intelligence*, 170(2):114–159, 2006.

[66] Phan Minh Dung and Tran Cao Son. An argument-based approach to reasoning with specificity. *Artificial Intelligence*, 133(1-2):35–85, 2001.

[67] Phan Minh Dung and Phan Minh Thang. Towards (probabilistic) argumentation for jury-based dispute resolution. In *Computational Models of Argument*, volume 216, pages 171–182, 2010.

[68] Paul E Dunne, Anthony Hunter, Peter McBurney, Simon Parsons, and Michael

Wooldridge. Weighted argument systems: Basic definitions, algorithms, and complexity results. *Artificial Intelligence*, 175(2):457–486, 2011.

[69] Marilyn Ford. System LS: A three-tiered nonmonotonic reasoning system. *Computational Intelligence*, 20(1):89–108, 2004.

[70] Dov M Gabbay. Semantics for higher level attacks in extended argumentation frames part 1: Overview. *Studia Logica*, 93(2-3):357, 2009.

[71] Dov M Gabbay and Odinaldo Rodrigues. Probabilistic argumentation: An equational approach. *Logica Universalis*, 9(3):345–382, 2015.

[72] Alejandro García and Guillermo R. Simari. Defeasible logic programming: An argumentative approach. *Theory and Practice of Logic Programming*, 4(1+2):95–138, 2004.

[73] P Geerts and D Vermeir. A nonmonotonic reasoning formalism using implicit specificity information. In *Proceeding of the Conference on Logic Programming and Nonmonotonic Reasoning*, volume 93, pages 380–396, 1991.

[74] Angelo Gilio, Niki Pfeifer, and Giuseppe Sanfilippo. Transitive reasoning with imprecise probabilities. In *Symbolic and Quantitative Approaches to Reasoning with Uncertainty - European Conference, ECSQARU*, pages 95–105, 2015.

[75] Angelo Gilio, Niki Pfeifer, and Giuseppe Sanfilippo. Transitivity in coherence-based probability logic. *Journal of Applied Logic*, 14:46–64, 2016.

[76] Mauro J. Gómez Lucero, Carlos I. Chesñevar, and Guillermo R. Simari. Modelling argument accrual in possibilistic defeasible logic programming. In Claudio Sossai and Gaetano Chemello, editors, *Symbolic and Quantitative Approaches to Reasoning with Uncertainty*, pages 131–143. Springer Berlin Heidelberg, 2009.

[77] T.F. Gordon, H. Prakken, and D. Walton. The Carneades model of argument and burden of proof. *Artificial Intelligence*, 171(10):875–896, 2007.

[78] Thomas Gordon and Douglas Walton. Formalizing balancing arguments. In *Proceedings of the 2016 Conference on Computational Models of Argument (COMMA 2016)*, pages 327–338. IOS Press, 2016.

[79] Matthias Grabmair, Thomas F Gordon, and Douglas Walton. Probabilistic semantics for the Carneades argument model using Bayesian networks. In *Computational Models of Argument*, pages 255 266, 2010.

[80] Adam Grove. Two modellings for theory change. *Journal of philosophical logic*, 17(2):157–170, 1988.

[81] Jörg Hansen. Prioritized conditional imperatives: problems and a new proposal. *Autonomous Agents and Multi-Agent Systems*, 17(1):11–35, 2008.

[82] Carl G Hempel. Maximal specificity and lawlikeness in probabilistic explanation. *Philosophy of Science*, 35(2):116–133, 1968.

[83] Jesse Heyninck and Christian Straßer. Relations between assumption-based approaches in nonmonotonic logic and formal argumentation. In Gabriele Kern-Isberner and Renata Wassermann, editors, *Proceedings of NMR2016*, pages 65–76, 2016.

[84] Jesse Heyninck and Christian Straßer. Revisiting unrestricted rebut and preferences in

structured argumentation. In *Proceedings of the 26th International Joint Conference on Artificial Intelligence*, pages 1088–1092. AAAI Press, 2017.

[85] Jesse Heyninck and Christian Straßer. A comparative study of assumption-based approaches to reasoning with priorities. In *Second Chinese Conference on Logic and Argumentation*, 2018.

[86] J. Horty. Some direct theories of nonmonotonic inheritance. In Dov Gabbay, Christopher Hogger, and John Robinson, editors, *Handbook of Logic in Artificial Intelligence and Logic Programming, Vol. 3: Nonmonotonic Reasoning and Uncertain Reasoning*, pages 111–187. Oxford University Press, 1994.

[87] John Horty. *Reasons as Defaults*. Oxford University Press, 2012.

[88] John F Horty. Argument construction and reinstatement in logics for defeasible reasoning. *Artificial intelligence and Law*, 9(1):1–28, 2001.

[89] Anthony Hunter. Some foundations for probabilistic abstract argumentation. In *Computational Models of Argument*, volume 245, pages 117–128, 2012.

[90] Anthony Hunter. A probabilistic approach to modelling uncertain logical arguments. *International Journal of Approximate Reasoning*, 54(1):47–81, 2013.

[91] Anthony Hunter and Matthias Thimm. Probabilistic reasoning with abstract argumentation frameworks. *Journal of Artificial Intelligence Research*, 59:565–611, 2017.

[92] Jeroen Janssen, Martine De Cock, and Dirk Vermeir. Fuzzy argumentation frameworks. In *Information Processing and Management of Uncertainty in Knowledge-based Systems*, pages 513–520, 2008.

[93] Ralph H. Johnson. *Manifest Rationality. A Pragmatic Theory of Argument*. Routledge, 2000.

[94] Souhila Kaci and Christophe Labreuche. Arguing with valued preference relations. In *European Conference on Symbolic and Quantitative Approaches to Reasoning and Uncertainty*, pages 62–73. Springer, 2011.

[95] Souhila Kaci and Leendert van der Torre. Preference-based argumentation: Arguments supporting multiple values. *International Journal of Approximate Reasoning*, 48(3):730–751, 2008.

[96] Jeroen Keppens. Argument diagram extraction from evidential Bayesian networks. *Artificial Intelligence and Law*, 20(2):109–143, 2012.

[97] John Maynard Keynes. *A treatise on probability*. London: Macmillan, 1921.

[98] Lewis A. Kornhauser. Modelling collegial courts. ii. legal doctrine. *Journal of Law, Economics and Organization*, 8:441–470, 1992.

[99] Sarit Kraus, Daniel Lehmann, and Menachem Magidor. Nonmonotonic reasoning, preferential models and cumulative logics. *Artificial intelligence*, 44(1-2):167–207, 1990.

[100] Henry E Kyburg Jr and Choh Man Teng. *Uncertain inference*. Cambridge University Press, 2001.

[101] Hengfei Li, Nir Oren, and Timothy J Norman. Probabilistic argumentation frameworks. In *International Workshop on Theorie and Applications of Formal Argumen-*

tation, pages 1–16. Springer, 2011.

[102] Zimi Li, Nir Oren, and Simon Parsons. On the links between argumentation-based reasoning and nonmonotonic reasoning. *arXiv preprint arXiv:1701.03714*, 2017.

[103] Beishui Liao, Nir Oren, Leon van der Torre, and Serena Villata. Prioritized norms and defaults in formal argumentation. *Deontic Logic and Normative Systems (2016)*, 2016.

[104] Ronald Prescott Loui. Defeat among arguments: a system of defeasible inference. *Computational intelligence*, 3(1):100–106, 1987.

[105] Mauro Javier Gómez Lucero, Carlos Iván Chesñevar, and Guillermo Ricardo Simari. On the accrual of arguments in defeasible logic programming. In *Proceedings of the International Joint Conference on Artificial Intelligence, IJCAI*, 2009.

[106] Diego C. Martínez, Alejandro Javier García, and Guillermo Ricardo Simari. On acceptability in abstract argumentation frameworks with an extended defeat relation. In Paul E. Dunne and Trevor Bench-Capon, editors, *Computational Models of Argument*, volume 144 of *Frontiers in Artificial Intelligence and Applications*, pages 273–278. IOS Press, 2006.

[107] Diego C. Martínez, Alejandro Javier García, and Guillermo Ricardo Simari. An abstract argumentation framework with varied-strength attacks. In *Proceedings of the International Conference on Principles of Knowledge Representation and Reasoning (KR)*, pages 135–144, 2008.

[108] Diego C. Martínez, Alejandro Javier García, and Guillermo Ricardo Simari. An abstract argumentation framework with varied-strength attacks. In Gerhard Brewka and Jérôme Lang, editors, *Proceedings of the International Conference on Principles of Knowledge Representation and Reasoning (KR)*, pages 135–144. AAAI Press, 2008.

[109] Paul-Amaury Matt and Francesca Toni. A game-theoretic measure of argument strength for abstract argumentation. In *European Workshop on Logics in Artificial Intelligence*, pages 285–297. Springer, 2008.

[110] Sanjay Modgil. Reasoning about preferences in argumentation frameworks. *Artificial intelligence*, 173(9-10):901–934, 2009.

[111] Sanjay Modgil and Trevor Bench-Capon. Integrating object and meta-level value based argumentation. In *Proceedings of the 2008 conference on Computational Models of Argument: Proceedings of COMMA 2008*, pages 240–251. IOS Press, 2008.

[112] Sanjay Modgil and Henry Prakken. A general account of argumentation with preferences. *Artificial Intelligence*, 195:361–397, 2013.

[113] Sanjay Modgil and Henry Prakken. The ASPIC+ framework for structured argumentation: a tutorial. *Argument & Computation*, 5(1):31–62, 2014.

[114] Sanjay Modgil and Henry Prakken. Abstract rule-based argumentation. In *Handbook of Formal Argumentation*. College Publications, 2018.

[115] Shyam Nair and John Horty. The logic of reasons. In D. Star, editor, *The Oxford Handbook of Reasons and Normativity*, pages 67–84. Oxford University Press, 2018.

[116] Rescher Nicholas. *Plausible reasoning. An introduction to the theory and practice of*

plausibilistic inference. BRILL, 1978.

[117] Søren Holbech Nielsen and Simon Parsons. A generalization of Dung's abstract frame-work for argumentation: Arguing with sets of attacking arguments. In *Argumentation in Multi-Agent Systems*, pages 54–73. Springer, 2006.

[118] Farid Nouioua. Afs with necessities: Further semantics and labelling characterization. In Weiru Liu, V. S. Subrahmanian, and Jef Wijsen, editors, *Scalable Uncertainty Management - 7th International Conference, SUM 2013, Washington, DC, USA, September 16-18, 2013. Proceedings*, volume 8078 of *Lecture Notes in Computer Science*, pages 120–133. Springer, 2013.

[119] Donald Nute. Defeasible logic. In D. Gabbay and C. Hogger, editors, *Handbook of Logic for Artificial Intelligence and Logic Programming*, volume III, pages 353–395. Oxford University Press, 1994.

[120] Nir Oren and Timothy J. Norman. Semantics for evidence-based argumentation. In *Computational Models of Argument*, pages 276–284, 2008.

[121] Célia Da Costa Pereira, Andrea GB Tettamanzi, and Serena Villata. Changing one's mind: erase or rewind? possibilistic belief revision with fuzzy argumentation based on trust. In *Proceedings of the International Joint Conference on Artificial Intelligence, IJCAI*, 2011.

[122] Sylwia Polberg, Anthony Hunter, and Matthias Thimm. Belief in attacks in epistemic probabilistic argumentation. In *International Conference on Scalable Uncertainty Management*, pages 223–236. Springer, 2017.

[123] Sylwia Polberg and Nir Oren. Revisiting support in abstract argumentation systems. In Simon Parsons, Nir Oren, Chris Reed, and Federico Cerutti, editors, *Computational Models of Argument*, volume 266 of *Frontiers in Artificial Intelligence and Applications*, pages 369–376. IOS Press, 2014.

[124] John Pollock. Epistemology and probability. *Synthese*, 55(2):231–252, 1983.

[125] John Pollock. Defeasible reasoning. *Cognitive Science*, 11(4):481–518, 1987.

[126] John Pollock. *Cognitive Carpentry. A Blueprint for How to Build a Person.* MIT Press, 1995.

[127] David Poole. On the comparison of theories: Preferring the most specific explanation. In *Proceedings of the International Joint Conference on Artificial Intelligence, IJCAI*, IJCAI'85, pages 144–147, San Francisco, CA, USA, 1985. Morgan Kaufmann Publishers Inc.

[128] Karl Popper. *The logic of scientific discovery.* London: Hutchinson, 1972.

[129] Henry Prakken. A study of accrual of arguments, with applications to evidential reasoning. In *Proceedings of the 10th International Conference on Artificial Intelligence and Law*, ICAIL '05, pages 85–94, New York, NY, USA, 2005. ACM.

[130] Henry Prakken. An abstract framework for argumentation with structured arguments. *Argument and Computation*, 1(2):93–124, 2010.

[131] Henry Prakken. On support relations in abstract argumentation as abstractions of inferential relations. In Torsten Schaub, Gerhard Friedrich, and Barry O'Sullivan, ed-

itors, *European Conference on Artificial Intelligence (ECAI)*, volume 263 of *Frontiers in Artificial Intelligence and Applications*, pages 735–740. IOS Press, 2014.

[132] Henry Prakken. On relating abstract and structured probabilistic argumentation: a case study. In *European Conference on Symbolic and Quantitative Approaches to Reasoning and Uncertainty*, pages 69–79. Springer, 2017.

[133] Antonio Rago, Francesca Toni, Marco Aurisicchio, Pietro Baroni, et al. Discontinuity-free decision support with quantitative argumentation debates. In *Proceedings of the International Conference on Principles of Knowledge Representation and Reasoning (KR)*, volume 16, pages 63–73. AAAI Press, 2016.

[134] Jussi Rintanen. Prioritized autoepistemic logic. In *European Workshop on Logics in Artificial Intelligence*, pages 232–246. Springer, 1994.

[135] Régis Riveret, Pietro Baroni, Yang Gao, Guido Governatori, Antonino Rotolo, and Giovanni Sartor. A labelling framework for probabilistic argumentation. *Annals of Mathematics and Artificial Intelligence*, 83(1):21–71, 2018.

[136] Régis Riveret, Antonino Rotolo, Giovanni Sartor, Roth Bram, and Henry Prakken. Success chances in argument games: a probabilistic approach to legal disputes. In *Legal Knowledge and Information Systems (JURIX'07)*, pages 99–108. IOS Press, 2007.

[137] Nico Roos. Preferential model semantics, argumentation frameworks and closure properties. In *Proceedings of the 14th International Workshop on Non-Monotonic Reasoning (NMR 2012)*, 2012.

[138] Claudia Schulz and Francesca Toni. Justifying answer sets using argumentation. *Theory and Practice of Logic Programming*, 16(1):59–110, 2016.

[139] Dunja Šešelja and Christian Straßer. Abstract argumentation and explanation applied to scientific debates. *Synthese*, 190(12):2195–2217, 2013.

[140] Wolfgang Spohn. Ordinal conditional functions: a dynamic theory of epistemic states. In William L. Harper and Brian Skyrms, editors, *Causation in Decision, Belief Change, and Statistics*, volume II, pages 105–134, 1988.

[141] Frieder Stolzenburg, Alejandro J García, Carlos I Chesnevar, and Guillermo R Simari. Computing generalized specificity. *Journal of Applied Non-Classical Logics*, 13(1):87–113, 2003.

[142] Nouredine Tamani and Madalina Croitoru. Fuzzy argumentation system for decision support. In Anne Laurent, Olivier Strauss, Bernadette Bouchon-Meunier, and Ronald R. Yager, editors, *Information Processing and Management of Uncertainty in Knowledge-Based Systems*, pages 77–86, Cham, 2014. Springer International Publishing.

[143] Nouredine Tamani and Madalina Croitoru. Fuzzy argumentation system for decision support. In *International Conference on Information Processing and Management of Uncertainty in Knowledge-Based Systems*, pages 77–86. Springer, 2014.

[144] Matthias Thimm. A probabilistic semantics for abstract argumentation. In *European Conference on Artificial Intelligence (ECAI)*, pages 750–755, 2012.

[145] Matthias Thimm and Gabriele Kern-Isberner. On controversiality of arguments and

stratified labelings. In *Computational Models of Argument*, pages 413–420, 2014.

[146] Sjoerd T Timmer, John-Jules Ch Meyer, Henry Prakken, Silja Renooij, and Bart Verheij. A two-phase method for extracting explanatory arguments from Bayesian networks. *International Journal of Approximate Reasoning*, 80:475–494, 2017.

[147] Francesca Toni. A tutorial on assumption-based argumentation. *A&C*, 5(1):89–117, 2014.

[148] Frederik Van De Putte and Christian Straßer. Three formats of prioritized adaptive logics: a comparative study. *Logic Journal of IGPL*, 21(2):127–159, 2012.

[149] B. Verheij. *Rules, Reasons, Arguments. Formal Studies of Argumentation and Defeat.* Dissertation, Maastricht University, 1996.

[150] Bart Verheij. Accrual of arguments in defeasible argumentation. In *In Proceedings of the Second Dutch/German Workshop on Nonmonotonic Reasoning*, pages 217–224, 1995.

[151] Bart Verheij. Arguments and their strength: Revisiting pollock's anti-probabilistic starting points. In *Computational Models of Argument*, pages 433–444, 2014.

[152] Bart Verheij. To catch a thief with and without numbers: arguments, scenarios and probabilities in evidential reasoning. *Law, Probability and Risk*, 13(3-4):307–325, 2014.

[153] Bart Verheij. Proof with and without probabilities. *Artificial Intelligence and Law*, 25(1):127–154, 2017.

[154] Ignacio Darío Viglizzo, Fernando A. Tohmé, and Guillermo Ricardo Simari. The foundations of delp: defeating relations, games and truth values. *Annals of Mathematics and Artificial Intelligence*, 57(2):181–204, 2009.

[155] Gerard AW Vreeswijk. Argumentation in Bayesian belief networks. In *International Workshop on Argumentation in Multi-Agent Systems*, pages 111–129. Springer, 2004.

[156] Toshiko Wakaki. Assumption-based argumentation equipped with preferences. In *International Conference on Principles and Practice of Multi-Agent Systems*, pages 116–132. Springer, 2014.

[157] Claus-Peter Wirth and Frieder Stolzenburg. A series of revisions of david poole's specificity. *Annals of Mathematics and Artificial Intelligence*, 78(3-4):205–258, 2016.

[158] Safa Yahi, Salem Benferhat, Sylvain Lagrue, Mariette Sérayet, and Odile Papini. A lexicographic inference for partially preordered belief bases. In *Proceedings of the International Conference on Principles of Knowledge Representation and Reasoning (KR)*, volume 8, pages 507–517. AAAI Press, 2008.

[159] Anthony P Young, Sanjay Modgil, and Odinaldo Rodrigues. Prioritised default logic as argumentation with partial order default priorities. *arXiv preprint arXiv:1609.05224*, 2016.

[160] Anthony P Young, Sanjay Modgil, and Odinaldo Rodrigues. Prioritised default logic as rational argumentation. In *Proceedings of the 2016 International Conference on Autonomous Agents & Multiagent Systems*, pages 626–634. International Foundation for Autonomous Agents and Multiagent Systems, 2016.

[161] Bruno Yun, Pierre Bisquert, Patrice Buche, and Madalina Croitoru. Arguing about

end-of-life of packagings: Preferences to the rescue. In *Research Conference on Metadata and Semantics Research*, pages 119–131. Springer, 2016.

[162] Bruno Yun, Madalina Croitoru, and Pierre Bisquert. Are ranking semantics sensitive to the notion of core? In *Proceedings of the 16th Conference on Autonomous Agents and Multiagent Systems*, pages 943–951. International Foundation for Autonomous Agents and Multiagent Systems, 2017.

Received 9 June 2018

Imprecise Probability and the Measurement of Keynes's "Weight of Arguments"

William Peden

Centre for Humanities Engaging Science and Society, 50 Old Elvet, DH1 3HN, Department of Philosophy, Durham University, UK

w.j.peden@durham.ac.uk

Abstract

Many philosophers argue that Keynes's concept of the "weight of arguments" is an important aspect of argument appraisal. The weight of an argument is the quantity of relevant evidence cited in the premises. However, this dimension of argumentation does not have a received method for formalisation. Kyburg has suggested a measure of weight that uses the degree of imprecision in his system of "Evidential Probability" to quantify weight. I develop and defend this approach to measuring weight. I illustrate the usefulness of this measure by employing it to develop an answer to Popper's Paradox of Ideal Evidence.

Introduction

One of the oldest systems of imprecise probability in the literature is Henry E. Kyburg's system of "Evidential Probability". In this article, I shall defend and develop a proposal by Kyburg to use his system to measure what John Maynard Keynes called the "weight of arguments" [1]. The analysis of weight has been challenging for probabilistic theories of reasoning, but I shall argue that Kyburg's measure can address this issue. My discussion will illustrate the advantages of Evidential Probability for argument analysis. Additionally, to illustrate the usefulness of this measure, I shall use this measure to answer Karl Popper's "Paradox of Ideal Evidence". While there are many proposed methods for measuring weight, I shall focus entirely on Kyburg's measure.

In Section 1, I describe Keynes's concept of weight and explain its importance. In Section 2, I explain Kyburg's system and develop his proposal for measuring weight using his system; I also examine whether this measure can withstand a range of existing and novel objections to using imprecise probabilities to measure weight. In Section 3, I apply this measure to Popper's paradox.

1 Keynes and the Weight of Argument

Keynes uses the term "weight of argument" to refer to his concept. In the literature, "weight of evidence" is also common. I shall stick to Keynes's terminology, because the phrase "weight of evidence" sometimes refers to other concepts, such as the degree of confirmation of a hypothesis by the evidence. Additionally, in law, the phrase "weight of evidence" is typically used to mean the balance of evidence with respect to a hypothesis of guilt or innocence [2, p. ix]. Thus, "weight of argument" helps avoid some potential ambiguities.

The "weight of argument" of a statement H given another statement E is the extent to which E (which might be the conjunction of many distinct statements in natural language) provides information that is relevant to H. As Rod O'Donnell emphasises, the adjective 'relevant' in front of 'evidence' is very important [3]. Keynes is not referring to the sheer number of statements on the right hand side of a conditional probability $P(H \mid E)$ or the sheer bulk of information that these statements contain. By "relevant evidence", Keynes is only referring to the extent that E provides information that is pertinent to H in particular.

The informal concept of weight has a long history. I. J. Good notes that, as a metaphor, it can be traced back at least to the Ancient Greek goddess Themis and her 'scales of justice', in which the scales involve both a balance and a quantity of evidence [4]. Charles Sanders Peirce provides one of the earliest philosophical discussions [5]. He gives a simple example of weight's significance: suppose there is a bag that we know consists of red beans and/or black beans, in some unknown proportion. Intuitively, there is a difference between extrapolating from half of a sample of 1,000 beans drawn from the bag and half of a sample of only 2 beans. If someone is extrapolating from the former bag, then they seem to be entitled to a greater degree of confidence (in some informal sense) in estimating that half of the beans of a bag are red. According to Peirce, the difference is the greater quantity of evidence that the report of the larger sample provides. Thus, an argument from that report and the relevant background knowledge to a hypothesis about the proportion of red beans in the bag has more weight.

Keynes firmly distinguishes weight and conditional probability, because they can move in different directions. For example, imagine that you hear a weather report stating that the weather in your area tomorrow will not be windy. The conditional probability that tomorrow *will* be windy, given your total evidence, has decreased. In contrast, the weight of argument has *increased*. Jochen Runde notes that Keynes presupposes that weight always increases monotonically with additions of relevant evidence to a body of evidence [6]. This monotonicity of weight means that, if E is relevant to H given background knowledge K, then $(E \wedge K)$ must have greater weight with respect to H than K alone. In this respect, weight differs from conditional probability, because conditional probability does not always increase monotonically as relevant statements are added to the conditional evidence.

However, some philosophers who have discussed weight have questioned this monotonicity presupposition. Runde argues that, under some circumstances, new evidence might reveal that our evidence is more limited than we thought [6]. James M. Joyce implicitly rejects the monotonicity presupposition in his discussion of weight [7]. Brian Weatherson argues against the monotonicity presupposition and provides an elegant example to support his denial [8]. Imagine that you are playing poker with several people. You are wondering if another player, whom I shall call "Lydia", has a straight flush. A straight flush in poker consists of five cards of sequential rank in the same suite. For instance, a hand consisting of a two, a three, a four, a five, and a six, all Diamonds, would be a straight flush. Since a dealt poker hand consists of 5 cards from a normal deck of 52 cards, there are $\frac{52!}{(5!)(47!)} = 2,598,960$ possible hands that Lydia might have. There are 10 possible straight flushes for each suit and 4 suits. Therefore, there are 40 possible straight flush. In the absence of any additional evidence that makes a particular hand more likely, the Keynesian probability that Lydia has a straight flush is $\frac{40}{2,598,960}$. In the Keynesian theory of evidence and probability, this initial total evidence provides substantial information about your conjecture that she has a straight flush. However, the subsequent combination of this initial evidence with Lydia's facial expressions, tone of voice, past bluffing behaviour, and other sources of information regarding her hand might leave you with much more vague total evidence. In Keynes's probability theory, this vagueness might make it impossible to specify a precise conditional probability for the conjecture that Lydia has a straight flush.

In Weatherson's example, the subsequent evidence is qualitative. However, there are also possible scenarios in which the evidence is entirely quantitative. For instance, imagine that I am about to make a selection from an urn containing 100 lottery tickets. 90 of the tickets are red and 10 are blue. You are wondering whether I shall draw a blue ticket. If there is no additional evidence that suggests that any one ticket is more likely to be selected than another, then the initial Keynesian probability of the hypothesis that I shall draw a blue ticket is $\frac{10}{100} = 0.1$. Suppose that you then learn that (1) I am drawing the ticket from the top of the urn and you also learn that (2) the proportion of blue tickets at the top of the urn is 0.2, 1, or somewhere in between. In other words, you now know that the ticket will be selected from a subset of the urn whose proportion of blue tickets does not match the proportion for the urn as a whole, but you only have imprecise statistical data about the subset: the proportion of blue tickets x has a value $0.2 \geq x \leq 1$. While you have acquired new evidence about the hypothesis that the ticket will be blue, it seems that the weight of your total evidence has been reduced, because the pertinent evidence of proportions has become imprecise.

Weatherson argues that, in such situations, it is intuitive to say that the weight of argument has fallen after you have acquired this new evidence. Metaphorically, Weatherson is pointing to the possibility that additional relevant information can result in a murkier overall picture, such that the total evidence provides vaguer information regarding a proposition and therefore less weight. On the grounds of such examples, I shall *not* share Keynes's presupposition (which he does not defend) that weight is monotonic.

Keynes did not doubt that weight was an important concept, but he was unsure about precisely *how* weight has practical significance. However, there have been a considerable number of suggested applications. In formal epistemology, Janina Hosiasson utilises Keynes's concept as part of an analysis of evidence and its value [9]. Weatherson argues that probabilistic theories of reasoning, such as Bayesianism, can incorporate weight to distinguish (a) propositions for which our total evidence provides considerable information and (b) propositions for which our total relevant evidence is highly exiguous, which is a distinction that is not captured by conditional probabilities [10]. In the philosophy of law, Barbara Davidson and Robert Pargetter use weight in their analysis of the legal phrase 'beyond reasonable doubt'[11]. James Franklin has also employed the concept in the philosophy of law [12,13]. These are just a few examples of how Keynes's concept has been fecund in its philosophical applications.

Keynes doubted that weight is quantitatively measurable. However, a quantitative formalization of weight might be useful, given the employment of the concept of weight by formal epistemologists and philosophers of law. Imprecise probabilities offer one possible basis for such measurement.

2 Evidential Probability and the Weight of Argument

The oldest suggestion of using imprecise probabilities to measure weight seems to be Kyburg's proposal in a 1961 monograph. He proposed using the degree of imprecision of the intervals in his probability system, called 'Evidential Probability' [14]. (I shall use the upper-case 'Evidential Probability' to refer to Kyburg's system and lower-case phrases like, 'evidential probability' for the probability values in his system.) The probabilities in this system take the values of intervals in the range from 0 to 1. For example, the intervals can cover the whole range, such as $[0, 1]$, or some subinterval of the range, such as $[0.5, 0.75]$, or a degenerate interval such as $[0.5, 0.5]$. In a 1968 article, Kyburg restricts his claim to the measurement of weight where the evidence is of the same sort [15]. His proposed measure is a very simple function:

Definition 1 *Weight of argument for H given E and K* $= (y - x)$ *where H is a hypothesis, E is some evidence, K is the relevant background information, x is the lower bound of the evidential probability interval for H given $(E \wedge K)$, and y is the upper bound of the interval for H given $(E \wedge K)$.*

However, this measure is linguistically awkward, since the values will be high when weight is low and low when weight is high. A simple modification makes the definition more verbally felicitous:

Definition 2 *Weight of argument for H given E and K* $= WK = 1 - (y - x)$

Thus, insofar as the Evidential Probability intervals are more imprecise (wider) the weight will be low; insofar as the intervals are precise, the weight will be high. Thus, when the value of H given $(E \wedge K)$ is a degenerate interval like $[0.5, 0.5]$, then the WK measure of weight will take a maximal value: $1 - (0.5 - 0.5) = 1$. When the interval is the maximally wide $[0, 1]$, the measure's value will take a minimal value: $1 - (1 - 0) = 0$. When the interval is a non-degenerate sub-interval, the measure's value will take some value between 1 and 0, depending on the imprecision of the evidential probability.

I shall argue that Kyburg's measure seems to offer a very widely applicable measure of weight, which goes at least beyond his 1968 claim. I shall begin by describing Kyburg's system of Evidential Probability. I shall then detail some simple examples of the WK measure in action. I shall finish this subsection by examining a range of existing and novel challenges for this measure.

2.1 Evidential Probability

Kyburg developed his system over 40 years, but I shall discuss his system in terms of its final version, as expressed in [16]. Kyburg's system of Evidential Probability maps interval values onto statements relative to other statements. Such values are intended to represent objective evidential relations between the two *relata*: the interval values are interpreted as a form of (non-deductive) logical relations, rather than in terms of belief. Thus, Evidential Probability is a member of the family of logical interpretations of probability. However, unlike many philosophers who use such an interpretation, Kyburg argues that the probabilities should be imprecise, in many contexts.

For the formal application of Evidential Probability, the statements in the probability relation are expressions within a formal language that models the reasoning of some agent(s). An Evidential Probability function EP has a value

$EP(H \mid E \wedge K) = [x, y]$, where the range of the function is interpreted as representing the objective degree of support that the conjunction of E and K provide for H and the interval is a closed interval with two real fraction as its limits.[1]

Instead of symmetry principles (as in Objective Bayesianism) or subjective credences (as in Subjective Bayesianism), Kyburg uses direct inference as the foundation of his probabilities. To use a simplified example, imagine that all you know about a single object a is that it is a fish. You also know that 1.5% of fish are goldfish. Kyburg (and many other philosophers) would say that the probability of the hypothesis Fa, given what you know, is 1.5%. However, this situation is extremely unrealistic. Our statistical data is typically imprecise: the proportions of predicates like 'fish' in populations are estimated with margins of error. And, as Hans Reichenbach argued [17], we always have competing statistics about reference classes for any given individual: we would not just know that a is a fish, but a fish that is differentiated from other fish via some definite description, such as 'That orange fish in Tank 7'. There is no reason why our data for the proportion of goldfish among orange fish or among fish in Tank 7 must match our data for fish in general. Thus, for Kyburg, the foundational issue in probability is formulating rules for ignoring and/or combining such competing reference class data.

To determine the range of the EP function for a hypothesis H and a body of evidence K, Kyburg proposes the following procedure. Where H concerns a single object a and a predicate F, one begins by enumerating the reference classes to which (i) a is known to belong and for which (ii) K contains data about the frequency of objects with F in that reference class. Such data can be imprecise: for example, K might include a statement R_i that only specifies that F has a relative frequency of 0.4 ± 0.05 in a particular reference class that includes the object a. For these reference classes, there will be a set of statements about frequencies in reference classes: $\{R_1, R_2, R_3 \ldots R_n\}$. These are the possible reference class statements, which report relative frequencies that might supply the interval values for $EP(H \mid K)$. Some of these reference class statements might *conflict*, in the sense that the relative frequency reported by one statement is not a subinterval of the relative frequency in the other: for example, perhaps you know that $1.5\% \pm 0.1\%$ of fish are goldfish, whereas you know that $45\% \pm 0.01\%$ of fish in Tank 7 are goldfish. Thus, there is a reference class problem: what is the appropriate reference class statement for the probability of a hypothesis, given competing reference class statements in K? Or, if there are no reasons to favour any single reference class statement to determine the

---

[1] I am expressing the evidential probabilities in a formalism that somewhat differs from Kyburg's, in that it is closer to the formalism of Bayesian conditional probabilities. This choice is to make it more familiar to readers who are unfamiliar with Evidential Probability. The differences from Kyburg are only superficial.

probability of the hypothesis, how should one combine conflicting statements into a single interval value?

Kyburg's answer is to sequentially apply the following rules to this set:

(1) Sharpening by Richness Compare each statement in the set via a pairwise comparison. Suppose that R_j has been deduced from a full joint distribution for a random variable Q, whereas R_i has been deduced from a marginal distribution for Q. Suppose that the interval of the relative frequency reported in R_j *conflicts* with the relative frequency reported in R_i. Sharpening by Richness requires that one ignores R_i. Put another way, Sharpening by Richness enables us to favour reference class statements that more fully articulate our statistical data.

For example, for a hypothesis about a selection from a set and given the choice between (1) statistical data about the *raw proportion* in the set and (2) statistical data about the *proportion of selections* from that set, one should favour the latter. Imagine that you are about to make a random selection from one of two decks of cards. You wondering whether to bet that the card you draw will be an Ace of Spades. Let H be the hypothesis that the card you will draw is the Ace of Spades. You know that Deck 1 is a normal deck and that Deck 2 is a normal deck *minus* the Spades. Before making your random selection, you must toss a six-sided die that you know to be normal and fair. If the die lands on 1 or 2, you must draw from Deck 1; if the die lands on a 3, 4, 5, or 6, you must draw from Deck 2. Thus, if the die lands on 2/3rds of the equiprobable possibilities, then you are certain to draw a card other than the Ace of Spades. Should you use (a) your information about the proportion of Aces of Spades in the total number of cards (1 out of 103) or (b) your information about the relative frequency of tossing the die *and* selecting the Ace of Spades? The former implies an evidential probability of $[\frac{1}{103}, \frac{1}{103}]$, whereas the latter joint distribution is the relative frequency of a 1 or 2 multiplied by the conditional frequency of selecting an Ace given that the die has landed favourably: $\left(\frac{1}{3}\right)\left(\frac{1}{52}\right) = \frac{1}{156}$, for an interval of $\left[\frac{1}{156}, \frac{1}{156}\right]$. Sharpening by Richness requires that you ignore the relative frequency about the cards alone, in favour of the joint distribution concerning the random variables of the card draw *and* the die toss. Assuming no other relative frequency statements survive competition against the latter data, the evidential probability of H will be $\left[\frac{1}{156}, \frac{1}{156}\right]$. In general, Sharpening by Richness means that statistical statements in the set of candidates for an evidential probability which are based on joint distributions must be favoured over those that are based on marginal distributions, in those circumstances where they conflict.

(2) **Sharpening by Specificity** Having applied the previous rule, one compares the surviving statements by a pairwise comparison. Suppose that the interval of the relative frequency reported in R_j conflicts with the relative frequency reported in R_i. If a statement R_j describes a proper subset of R_i, then R_j is preferred over R_i. Therefore, data that is more specifically about the individual in question will be favoured over more general conflicting data.

For example, suppose you are wondering if a flower in your greenhouse will bloom in March. If you know the general relative frequency of this species blooming in March and the more specific data for this species blooming in March when stored in a greenhouse, and these frequencies conflict, then one should ignore the data about the species in general. Similarly, in the earlier example of wondering if a is a goldfish, given choice between data for the proportion of goldfish among fish in general and data for the proportion of goldfish among fish in Tank 7 (the container of a) Sharpening by Specificity requires that one use the latter statistic, because you know that all the fish in Tank 7 are fish, but not all fish are in Tank 7. (The fish in Tank 7 are a proper subset of fish.) To use a gambling example, if you are selecting a ball from the upper region of an urn and you have data for the proportion of red balls on the top of the urn and for the proportion for the urn in general, then you should use your data for the upper region when determining the evidential probability that the ball is red.

Suppose that you have applied these two rules. If a single reference class statement R_j remains after one or both of these rules have been applied, then the evidential probability is $[x, y]$, where x and y are the lower and upper limits of the relative frequency data stated in R_j. For example, if R_j states that the relative frequency of F in a reference class is 0.5 ± 0.03, then $EP(Fa \mid K) = [0.47, 0.53]$. If a set of reference class statements remain, then the following rule must be applied:

(3) **Sharpening by Precision** If there is a statement R_j that states a relative frequency that is a proper subinterval of every other remaining statements' relative frequency data, then the evidential probability is this proper subinterval. If there is no such statement, then the evidential probability is the shortest possible cover of the intervals from the surviving members of the set.

For example, if R_1, R_2, and R_3 are the surviving statements and they provide intervals of $[0.1, 0.2]$, $[0.15, 0.3]$, and $[0.5, 0.55]$, then the evidential probability is $[0.1, 0.55]$.

Essentially the same reasoning for single-case probabilities extends to plural-case probabilities, because the Sharpening rules can be multi-member sets (a sample of balls from the top of an urn; the penguins of Antarctica; the stars in the Known Universe etc.) as well as hypotheses about particular individuals. Kyburg develops a range of general formal properties for values of the EP function: see [16,18], and a summary of salient results on page 49 in [19]. One axiom, that will be particular important later (in Subsection 2.2) is that if $EP(\Phi \leftrightarrow \Psi \mid \Omega, 1]$, then $EP(\Phi \mid \Omega) = EP(\Psi \mid \Omega)$, for any statements Φ, Ψ, and Ω. This has the consequence that if one knows that one statement is true if and only if a second statement is true, then the two statements must have the same evidential probability relative to one's total evidence.

Despite Kyburg's references to relative frequencies, he does not identify probabilities with relative frequencies: evidential probabilities take their values from accepted statements about frequencies, but they are not identified with frequencies. Instead, they are logical probabilities, akin to those of Keynes in [2] and Rudolf Carnap in [20]. As a result, he does not deny that probabilities can be meaningfully ascribed to single-case hypotheses, such as the hypothesis that a die will land on 6 in a particular toss. This feature of Evidential Probability enables Kyburg to obviate the standard frequentist problems with single-case probabilities. Indeed, as described above, all evidential probabilities are either single-case or ultimately derived from single-case probabilities.

Since Kyburg's proposal for measuring weight is my focus in this article, I shall not discuss the various general strengths and weaknesses of Kyburg's system. (The special issue on Kyburg in [21] contains further discussion, including critical assessments of Evidential Probability from Bayesian perspectives.) Instead, I shall restrict the scope my discussion to his measure of weight.

2.2 Examples

In Section 1, I mentioned a simple type of example of weight in reasoning, which was developed by Peirce. Suppose that you are considering hypotheses about the composition of an urn full of beans. You know that the urn consists of red and/or black beans in some unknown proportion. Assume that your background information is suitable for standard statistical inferences involving sampling without replacement: the selections are apparently random, they are not independent in your model of the sampling set-up, the population of beans is finite with well-defined means for the proportions of red beans and black beans, and so on. Intuitively, as your sample of beans increases, the quantity of evidence that is available to you (the weight) increases, *ceteris paribus*. I shall now explain how the WK measure treats this scenario.

Kyburg discusses this type of sample-to-population inferences in his system at great length in Chapter 11 of [16]. Since my focus is the WK measure, rather than the details of statistical inferences in his system, I shall be brief in my description of how such extrapolations work in Evidential Probability. Let H be the hypothesis that 49.5-50.5% of the beans in the urn are red. Let E_1 be the sample report that 2 sampled beans are red. Let E_2 be the sample report that 3,000 beans, including the beans described in E_1, also are red. Let K be your relevant background knowledge. Assume that there is sufficient information in K to calculate, using combinatorics, that between 2% and 100% of the 2-fold samples of any large finite population of beans in the urn will be matching samples, in the sense that the samples will match the population within a margin of error of 1%. The Law of Large Numbers is one obvious source of combinatoric information that might be useful, under suitable conditions, for calculating the proportion of matching samples. However, you might know other combinatoric principles that could help with the estimation; additionally, background information about the urn, urns in general, and other contingent facts might be relevant to the estimation of the proportion of matching samples in the urn. In short, the probability of an interval-valued hypothesis for a population, given some sample data and background information, will depend on the statistical estimate of matching samples of that size for that population, relative to the background information.

In this particular scenario, we have assumed that you know that the proportion of possible samples that match the population in redness, within a margin of error of 1%, is in the closed interval [0.02, 1]. If the rules of Sharpening select this statement about the set of possible compositions of the beans in the urn as your best information about H given the conjunction of E_1 and K, then the evidential probability will be $EP(H \mid E_1 \wedge K) = [0.02, 1]$. The WK measure gives the output of $(1 - (1 - 0.02)) = 0.02$ for the weight of E_1 and K with respect to H, which reflects Peirce's judgement that E_1 provides *some* evidence about H, but almost none.

In contrast, suppose that you can calculate that the relative frequency of 3,000-fold samples that match (within a margin of error of 1%) any large finite population of beans in the urn is some proportion from 72.665% to 100%. Assume that the conditions match those described in the paragraph, *mutatis mutandis*, such that $EP(H \mid E_1 \wedge E_2 \wedge K) = [0.72665, 1]$. The WK measure for your total evidence with respect to H is now $(1 - (1 - 0.72665)) = 0.72665$. In accordance with the intuition that Peirce is trying to convey, the WK measure gives us the result that your total evidence is much greater given a 3,000-fold sample report than a 2-fold sample report. In general, in the extremely simple sort of sampling model that Peirce is implicitly describing, if the evidence simply consists of increasingly larger sample reports, and if the evidential probabilities are simply derived from the combinatoric properties of possible samples of a large finite population, then the evidential probability intervals will narrow as the reported samples are larger. Thus, on the WK measure, the weight will increase in a linear fashion in such (idealized) circumstances.

I shall now give a different example in which the WK measure seems to work well. Imagine two different object-tossing scenarios: one in which the object you are tossing is an ordinary 1 coin; another in which the object you are tossing is a gömböc. A gömböc is a homogenous three-dimensional solid that has just two equilibria on a flat surface: (1) a stable equilibrium and (2) an unstable equilibrium. The gömböc that you are tossing has the Coptic letter '𝕎' on the side with its stable equilibrium point and the Coptic letter '𝐗' on the side with its unstable equilibrium point. You are wondering whether the next toss of the gömböc on a flat table in front of you will land '𝕎'. In the coin tossing case, you are wondering whether the next toss of the coin will land 'heads'. You might know that 1 coins land 'heads' with a relative frequency somewhere between 49% and 51%. Assume that, given your total evidence, this data about 1 coins in general is your best information (according to the rules of Sharpening) about whether the next toss will land 'heads', such that the evidential probability interval is [0.49, 0.51]. The weight of your total evidence, on the WK measure, is the very high value of $(1 - (0.51 - 0.49)) = 0.98$.

In contrast, imagine that you are almost entirely unfamiliar with the dynamics of the gömböc and gömböcs in general. Since gömböcs are rare (the first was produced in 2007 and the shape itself was only first conjectured in 1995 by the mathematician Vladimir Arnold) there might be little readily available information on how frequently they end up on any particular side. Nonetheless, imagine that you are able to conduct a brief investigation and learn that they land on their stable equilibrium point in somewhere between 0.01% and 99.9% of tosses on a flat table. Assume that this is your best information according to the rules of Sharpening. The evidential probability that the gömböc will land '𝕎', given your total evidence, is [0.01, 0.999]. The weight of your total evidence, on the WK measure, is the very low value of $(1 - (0.999 - 0.01)) = 0.011$. This value reflects the intuition that you have a much greater quantity of evidence about the hypothesis that the 1 coins will land 'heads' than the hypothesis that the gömböc will land '𝕎'.

These examples suggest that Kyburg's measure works well, at least in some simple cases of reasoning. There are many more complex cases where a measure of weight might be useful, and it would be preferable to have a broad survey of them. Yet such a survey would require a very wide-ranging analysis of many types of reasoning using Evidential Probability. In its place, I shall look at some *prima facie* problems for measures of weight that use imprecise probabilities, and examine whether the WK measure is vulnerable to them.

2.3 Dilation

Some imprecise probability systems have a feature called "dilation", in which updating on apparently irrelevant evidence can cause the probabilities to become more imprecise. Some philosophers try to use dilation as a general objection against such probability systems [22] and others have sought to defend imprecise probability systems by responding to such objections [23]. For my purposes in this article, the significant feature of dilation is the challenge that they can pose to measures of weight such as the WK measure. Seamus Bradley has recently raised such an objection: in dilation scenarios, the weight of argument for a hypothesis given the total evidence will fall upon the addition of apparently irrelevant evidence; consequently, a measure that identifies weight with the degree of imprecision will register a decrease upon the addition of apparently irrelevant evidence [24]. This raises the question of whether dilation is a feature of Evidential Probability. I shall set aside both (a) whether dilation is problematic and (b) whether dilation is problematic for all the types of weight measures that Bradley discusses. I shall focus purely on the WK measure. Arthur Paul Pederson and Gregory Wheeler [14], as well as Teddy Seidenfeld [25], have noted that dilation is not a feature of Evidential Probability, but they do not argue for this point in detail. In this subsection, I shall explain why the standard dilation scenarios are impossible in Kyburg's system.

I shall adapt Bradley's dilation scenario to Evidential Probability. Imagine that I am about to randomly select a card from a pile of 40 cards. Let H be 'The card is red', X be 'The card is even', Y be 'The card is red if and only if it is even', and K be my total evidence. Suppose that my best reference class information for H's probability is that $1/2$ of the cards in the pile are red, whereas I only know that the proportion of even-numbered cards in the pile is between 0 and 1. In some imprecise probability systems, learning Y results in a widening of the intervals of the probability of X given Y and K, in comparison to the probability of X given K alone.

In Evidential Probability, the value of $EP(X \mid Y \wedge K)$ is determined by the rules of Sharpening to the relative frequency data about X that $(Y \wedge K)$ implies. By assumption, the information about the relative frequency of cards in the pile is initially my best information about H. Since this data tells me that exactly half of the cards are red, such that $EP(H \mid K) = [0.5, 0.5]$. Learning Y gives me a new potential reference class statement: I now know that the card is a member of a subset of the cards in the pile (the subset of cards that are red if and only if they are even) and I must now apply the rules of Sharpening to check if I should favour my relative frequency data about this reference class (the data that states that the proportion of red cards in this class is between 0 and 1) over my information for the pile as a whole.

The intervals [0.5, 0.5] and [0, 1] do not conflict, in the sense described in Subsection 2.1, because [0.5, 0.5] is a proper subinterval of [0, 1]. Therefore, neither Sharpening by Richness nor Sharpening by Specificity is applicable. (The latter rule is not applicable, even though [0, 1] is derived from the evidence that the card in question is a member of a proper subset of the pile, because the intervals do not conflict and Sharpening by Specificity only favours more specific reference classes when they conflict with more general reference classes.) The remaining rule is Sharpening by Precision, which requires selecting a proper subinterval of the possible intervals that have survived the first two rules, *if* such a subinterval is available; otherwise, the evidential probability is the cover of the intervals. Since [0.5, 0.5] is a proper subinterval of [0, 1], Sharpening by Precision requires selecting [0.5, 0.5] as the value of $EP(H \mid Y \wedge K)$. Therefore, the Evidential Probability of H does not become more imprecise upon learning Y. The WK measure provides the output that learning Y has not affected the weight, because $WK(H \mid K) = 1 - (0.5 - 0.5) = 1$ and $WK(H \mid Y \wedge K) = 1 - (0.5 - 0.5) = 1$.

Additionally, Evidential Probability provides the potentially intuitive result that the weight of argument for X given Y and K is greater than the weight for H given K alone. As I noted towards the end of Subsection 2.2, if $EP(\Phi \leftrightarrow \Psi \mid \Omega) = [1, 1]$, then $EP(\Phi \mid \Omega) = EP(\Psi \mid \Omega)$, for any Φ, Ψ, and total evidence Ω. Therefore, since $EP(H \leftrightarrow X \mid (H \leftrightarrow X) \wedge K) = [1, 1]$, it follows that $EP(X \mid (H \leftrightarrow X) \wedge K) = EP(H \mid (H \leftrightarrow X) \wedge K)$. Using the reasoning in the earlier paragraph and the fact that Y is equivalent to $(H \leftrightarrow X)$, we can infer that the evidential probability that the card is even, given my new total evidence, must be [0.5, 0.5]. We can also infer this directly: the card is now known to be red if and only if it is even, such that the earlier [0, 1] interval is now competing with the [0.5, 0.5] interval from the data for red cards in the pile. Neither interval conflicts, because [0.5, 0.5] is a proper subinterval of [0, 1], and therefore Sharpening by Richness and Sharpening by Specificity are both inapplicable. Once again, via Sharpening by Precision, the new evidential probability must be [0.5, 0.5], because this interval is a proper subinterval of [0, 1]. Learning Y has increased the weight for X according to Kyburg's measure, because $WK(X \mid K) = 1 - (1 - 0) = 0$ and $WK(X \mid Y \wedge K) = 1 - (0.5 - 0.5) = 1$.

For some, these are both intuitive results. The WK measure tracks the intuition that Y adds to the weight for X and does not reduce the weight for H. However, those who believe that the imprecise probabilities *should* become more imprecise in dilation scenarios and/or that the weight should decrease will not share this intuition. To comprehensively defend the result in this subsection in depth would require a full defence of Evidential Probability against alternative approaches, and this defence would be outside the scope of my argument. However, there are two arguments that can be made for this output of Evidential Probability. Firstly, learning Y tells us that the card is a member of a subset of the pile, but it is a subset for which I have no (non-vacuous) data in Bradley's scenario; for all I know, the relative frequency of red cards in this subset is no different from the relative frequency of cards in the pile as a whole. In contrast, I *do* know that all the members of the subset are members of the pile and I have no reason to believe that they are unrepresentative of the pile with respect to their colour. As I do not seem to have any reasons to disregard the data for the pile, I should still use it.

Secondly, the claim that one should always use statistics for the most precise reference class has an apparent counterexample of singleton sets. I always know that the card is a member of the singleton containing that specific card. In Bradley's scenario, I am ignorant about the relative frequency of redness in this singleton. (Otherwise, I would have no reason to consider the pile as a whole, the subset, or any other broader reference class.) However, it seems unreasonable to say that, until I know whether the card is red or not red, my evidential probability for the hypothesis that it is red must be [0, 1], even though the singleton is the most specific possible reference class. This example suggests that, when conjecturing that a particular object a satisfies a predicate F, the mere fact that a is a member of a reference class A and that A is a proper subset of B is insufficient to require that the data for A must be favoured over the data for B. Kyburg avoids this consequence by limiting Sharpening by Specificity to comparisons in which two possible intervals are in conflict. Therefore, the evidential probability for a single-case hypothesis will be derived from the data about a singleton set *if* that data conflicts with any other candidates. For instance, if I know that the card is red, then the [1, 1] interval from my data for the singleton set will be selected over the [0.5, 0.5] interval from my data for the pile. However, such data must be ignored in Kyburg's system in cases where it does not conflict with more precise general data, such that the [0, 1] interval will not always dominate alternative data. Thus, in some circumstances and within the context of Kyburg's interpretation of probability (where probabilities are always derived via relative frequency data) it seems reasonable for relatively non-specific relative frequency information to be the basis of a probability. The selections of the [0.5, 0.5] value in the dilation scenario discussed above are instances of Kyburg's rules for adjudicating it is reasonable to use data for a reference class that is not the most specific.

One might wonder whether the evidential probabilities will widen in some variations on Bradley's scenario. For example, suppose that I learn that the card is a member of a subset of cards from a second pile. I do not know the proportion of red cards in this second pile. Will the evidential probability that the card is red, given my new total evidence, have a wider interval? If I am selecting from this subset, then some of my original evidence be contracted, because I have learned that it is false that the card will be selected from the first pile.[2] Suppose that (a) my interval for this subset is imprecise and (b) according to the rules of Sharpening, this interval must be selected over any competing possible interval. Under such circumstances, the evidential probability for the hypothesis that the card is red will widen. Yet, while this scenario exemplifies how evidential probabilities can become imprecise upon learning new data, it is not analogous to dilation: I acquired evidence that some of my earlier information was false, and (in contrast to scenarios featuring dilation) no-one thinks that such evidence is irrelevant. The WK measure will give a relatively imprecise value in this scenario, but that is an unsurprising feature, because we have assumed that the relatively precise frequency data has become unavailable.

However, while the classic form of dilation does not occur in Evidential Probability, there are circumstances in which learning new evidence can cause a widening of evidential probability intervals. Seidenfeld examines such a scenario in [22] and I shall now discuss this possibility.

2.4   The Hollow Cube

Seidenfeld developed his "Hollow Cube" scenario in order to demonstrate that Evidential Probabilists cannot always make use of Bayesian statistical methods. In particular, joint distributions of frequencies can only be utilised for determining evidential probabilities when the evidence contains those joint distributions. In some circumstances, this inability to feature of Evidential Probability results in the intervals widening. In Seidenfeld's scenario, we are measuring the volume of a hollow cube. We hypothesise that the cube has a volume of V millilitres, where V is an interval-value. Assume that we have two available measurement methods:

- (1) We can fill the cube with a liquid. We can then calculate the probability that the cube will have a volume V given that it has been filled by the measured quantity of that liquid.

- (2) We can cut a rod which has a length equal to the cube's edge and measure the length of this rod. We can then calculate the probability that the cube will have a volume V given the result of this cutting and measuring procedure.

---

[2]If I am not selecting from this subset and my data for the subset is relatively imprecise, then Sharpening by Precision will require ignoring the subset data in favour of the data for the pile as a whole.

Seidenfeld notes that Bayesians can always combine these results to calculate a posterior probability for the hypothesis: it is simply a matter of using the relevant priors and likelihoods from our existing full distribution to calculate the conditional probability of the hypothesis given the conjunction of the measurements. (This assumes, of course, that we happen to share the relevant Bayesian probabilities.) In contrast, using Evidential Probability, we can only use these Bayesian methods if we know a joint distribution for the frequency that the hypothesis will be true given the results of *both* measurements. However, Seidenfeld notes that such rich information might be unavailable. Potentially, we might just have the separate frequencies for the hypothesis's truth for each measurement; these frequencies might be very different. If the report of (2) is added after the report of (1) or *vice versa*, then the evidential probability can become wider after acquiring more evidence. As a result, the degree of imprecision will increase, and the WK measure decrease, such that learning the report of (2) results in less weight for an argument for/against the hypothesis given our total evidence.

For example, it is possible that using method (1) produces a measurement that strongly indicates that the cube's volume is in the interval V, whereas using method (2) produces a measurement that strong indicates that the volume is *not* in the interval V. I shall use the following abbreviations: E_1 is the estimate for the cube's volume given method (1), E_2 is the corresponding estimate given method (2), and H is the hypothesis that the volume of the cube is in the interval V. Our background knowledge is represented by K, and by assumption K lacks a sufficiently rich full distribution to use a joint distribution for H given E_1 and E_2 for Bayesian statistical methods. However, it does contain relative frequency data for the conditional probabilities of H given E_1 and H given E_2.

Since it is assumed that the joint distribution is not available, Evidential Probability will require using confidence-interval methods from classical statistics (if possible) as an alternative [16]. Suppose that we are using a confidence level such that the inference that $\neg H$ has a $\pm$ 2% confidence interval. Assume that H is extremely unlikely given E_1 and that $EP(H \mid E_1 \wedge K) = [0.01, 0.03]$. With comparable assumptions, suppose that E_2 provides an evidential probability that is almost a mirror image, such that H is very likely given our measurement using the rod: $EP(H \mid E_2 \wedge K) = [0.97, 0.99]$.

Sharpening by Richness is not available in this case, since we have assumed that there is no available joint distribution that provides frequency data for H given both measurements in our background knowledge. Sharpening by Specificity is not available, since neither $EP(H \mid E_1 \wedge K)$ nor $EP(H \mid E_2 \wedge K)$ is based on a reference class that we know to be the subset of the other. Only Sharpening by Precision remains. Since there is no available proper subinterval, this rule requires that we use the cover of the two potential interval, such that $EP(H \mid E_1 \wedge E_2 \wedge K) = [0.01, 0.99]$. Therefore, if we learned E_1 and obtained [0.01, 0.03] as the probability of H, then subsequently learning E_2 would result in a wider probability.

Kyburg accepts Seidenfeld's example without objection [26]. In itself, Seidenfeld's example is not problematic for Evidential Probability: if we discover that our initial probability for H was radically dependent on our choice of measurement method, then a less precise interval seems to provide an appropriate representation of our greater uncertainty upon learning this dependence. It also does not present a direct challenge to Kyburg's 1968 claim about the possible use of evidential probabilities to measure weight, because it involves different kinds of evidence; in his 1968 article, Kyburg only claims that the imprecision of evidential probabilities can measure the weight when the evidence is "of the same sort". Yet if one shares Keynes's assumption that weight must increase monotonically, Seidenfeld's scenario implies that the WK measure does not always capture the general informal concept of weight.

However, once we drop this monotonicity assumption, the Hollow Cube is no longer a problem for the WK measure. What is happening in Seidenfeld's scenario is analogous to Weatherson's poker example, which I discussed in Section 1. Adding new information to our total evidence about a hypothesis can result in a new body of evidence that is less informative about that hypothesis. Put figuratively, the combination of relevant evidence might have less weight than the sum of its parts. For instance, if we were estimating appropriate betting odds that the volume of a cube is within the interval V, then one might think that the range of reasonable betting odds is greater once we know the result of the second measurement. In other words, there is a greater degree of arbitrariness in any such decision according to Evidential Probabilists. (See Chapter 14 of [19] for Kyburg's account of how this idea can be explicated in his system and how evidential probabilities can guide betting decisions.) Using the WK measure, one can register this shift in the extent to which the evidence provides information once the discordance between the results of the two measurements becomes apparent.

One might object that a report of the second measurement is intuitively relevant to one's hypothesis, and therefore that the weight (which is the quantity of evidence in one's total information with respect to a hypothesis) should be greater. However, there is an ambiguity in "quantity of evidence" here. This phrase might mean (1) the quantity of relevant statements that are available to us, which *has* increased once the second measurement is reported. Yet it might also mean (2) the extent to which the *conjunction* of our total data (after learning the result of the second measurement) provides evidence about the hypothesis. In Seidenfeld's scenario, this latter quantity seems to have fallen, because the evidence is more ambiguous: the two measurements gave very different results, and we lack the background information to amalgamate them via Bayesian statistical methods. It is meaning (2) that Keynes meant by "weight". The WK measure formalises this fall in this quantity.

In general, the WK measure seems to work well even in those cases where evidential probabilities widen upon learning new evidence. Firstly, learning new evidence can cause one to abandon old evidence. For example, imagine that you are studying the empirical basis for a particular homeopathic medicine's efficacy. You learn that the putative studies of this medicine are fraudulent, because the tests never actually occurred. It is possible that your subsequent evidential probabilities for hypotheses like "This homeopathic treatment alleviates nasal congestion symptoms in 83–88% of adults" will be much wider due to the contraction of relative frequency data on the medicine's efficacy. The quantity of evidence has decreased in this scenario, even though the information that the results were fraudulent is relevant evidence for such hypotheses, and this decrease will be reflected in higher values of the WK measure.

Secondly, Sharpening by Richness and Sharpening by Specificity can require that relatively imprecise intervals must be used after new evidence is acquired. For instance, recall the card selection set-up discussed in Subsection 2.2: I am making a random selection from a pile of 40 cards and considering the hypothesis that the selected card will be red. Imagine that I learn that the card is from the top of the pile. Suppose that I know that 10–20% of cards at the top of the pile are red. The [0.1, 0.2] interval conflicts with the [0.5, 0.5] interval that I could derive from my data about the pile. Sharpening by Specificity requires that I use the wider interval. Therefore, the WK measure will indicate a loss of weight. However, this fall in the WK measure reflects the intuition that (a) I have learned that the card belongs to an unrepresentative subset of the pile and (b) my evidence about this subset is less precise than my data for the pile. By contrast, in the version of this scenario that I discussed in Subsection 2.2, condition (a) was not satisfied, since I did not know that the subset is unrepresentative of the pile in its colour. Assuming that conditions (a) and (b) entail a loss of weight, the WK measure provides an adequate analysis of this possibility.

These particular examples do not provide a general proof that there are no circumstances in which evidential probabilities widen and yet the weight does not seem to decrease. Since the intervals in Kyburg's system can widen under a variety of possible situations, there is always the possibility that such a widening might not correspond to the informal concept of weight. Nonetheless, if Kyburg's system is a generally adequate theory of epistemic probability (and I have not argued, in this article, that this claim is true) the intervals for a hypothesis given new total evidence should not widen or contract except via some important change in the total evidence. Furthermore, they suggest that the WK measure is widely applicable, even if it cannot address every possible circumstance in which we want to formalise weight.

2.5   The Problem of Corroborating Evidence

Evidential Probability intervals can be maximally imprecise: there is no wider value than $[0, 1]$ in Kyburg's formalism. Concomitantly, the WK measure can have a minimal value: when $EP(\Phi \mid \Omega) = [0, 1]$, then $WK(\Phi \mid \Omega) = (1 - (1 - 0)) = 0$. Evidential Probability intervals can also be maximally precise: if the intervals are degenerate, then no additional evidence can increase their precision. This entails that the WK measure can have a maximal value: when $EP(\Phi \mid \Omega) = [x, y]$ and $x = y$, then $WK(\Phi \mid \Omega) = (1 - (y - x)) = 1$. The Problem of Corroborating Evidence is that even once a hypothesis is inconsistent with our total evidence, such that its evidential probability is the degenerate interval $[0, 0]$, it seems possible to corroborate the evidence that is inconsistent with that hypothesis. Thus, the quantity of relevant evidence seems to be increasing while the WK measure stays at 1. This is a *prima facie* problem for the WK measure.

For instance, imagine that we are hypothesising that "All metal rods expand when they are heated" and that we are testing this hypothesis in a laboratory. In an experiment, we apply a heat source to the rod, but we compress the rod as it is heated. Let H be "All metal rods expand when heated", E_1 be "A metal rod was heated but did not expand" and E_2 be "A different metal rod was heated but did not expand." Suppose we accept E_1 after performing the experiment many times. (Due to measurement error and the fallibility of any instruments we could use, a single experiment will not be sufficient to establish E_1.) Since E_1 is inconsistent with H, our acceptance of E_1 reduces the evidential probability of H given the total evidence to $[0, 0]$, such that $WK(H \mid E_1 \wedge K) = 1$. This is because we know that the hypothesis is a member of the reference class of universal generalisations and we know from deductive logic that 0% of universal generalisations with counterexamples are true. Subsequently learning E_2 and conjoining it with $(E_1 \wedge K)$ will not provide an interval other than $[0, 0]$. Therefore, the WK measure value for H given $E_1 \wedge E_2 \wedge K)$ will not be different from 1. Yet one might think that our total evidence with respect to H has increased.

It is important to note that there is only a problem if E_1 has been fully accepted as evidence. This does not require that we regard E_1 as irrefutable, but it does mean that if we are Evidential Probabilists, then we must accept its immediate deductive consequences, including $\neg H$. If E_1 and E_2 merely become more probable at each stage, then the evidential probability of H relative to our total evidence will just become a lower and narrower interval, such that the WK measure increases. If accepting E_2 is genuinely adding apparently relevant information about H beyond the acceptance of E_1, then it must be via increasing the apparent reliability of E_1 and any other accepted statements that are inconsistent with H, *without* affecting the probability of H. However, the weight that a body of evidence supplies and the apparent reliability of that body of evidence are two different aspects of the strength of arguments. The WK measure is only intended to explicate weight; there is no apparent reason to require that it *also* captures apparent reliability.[3]

Once the distinction between the reliability of evidence and the quantity of evidence is made, the Problem of Corroborating Evidence does not pose a problem for the WK measure. However, it does raise the question of the relationship between weight and evidential relevance. It also illustrates that the "relevance" of evidence can be ambiguous. I shall discuss these issues later, in Section 3.

2.6 Inertia

As noted in the previous subsection, the WK measure will indicate a minimum weight when the Evidential Probability interval is the maximally wide [0, 1] interval. For example, imagine that there is a card that will be randomly drawn from a deck of unknown composition. Suppose that the only information provided by your total evidence, K, about the card's redness is that it will be drawn from a deck of cards and either 0%, 100%, or some intermediate proportion of the cards in this deck are red. Let H be the hypothesis that the card is red. Therefore, by the rules of Sharpening, $EP(H \mid K) = [0, 1]$, and the WK measure gives a value of 0 for the weight in this example. This illustrates the potential usefulness of the [0, 1] interval to enable the representation of minimal weight.

However, for some imprecise probability updating rules, it is not possible for additions of apparently relevant evidence to shift from the [0, 1] interval, unless either (a) the hypothesis has a probability of 1 given the new total evidence or (b) the hypothesis has a probability of 0 given the evidence. This feature, called "inertia", seems to have been first discovered by Peter Walley [30]. For the WK measure, the potential problem is that inertia could result in the measure failing to register apparent increases of evidence when (a) and (b) are not satisfied. For example, in the card selection scenario from the previous paragraph, suppose you learn the statement E, that '99% of the cards in the deck are red'. E seems to be relevant to H, but adding it to the total evidence will not change H's probability in some imprecise probability systems. Thus, if inertia is a feature of Evidential

---

[3]Kyburg does have an explication of the reliability of evidence, which he develops and applies in [16], [18], and [19].

Probability, then the WK measure would fail to register an apparent increase in weight.

As Isaac Levi has noticed, Evidential Probability does *not* feature inertia [31]. To illustrate this claim, I shall explain what the rules of Sharpening require when E is added to K. Since [0.5, 0.5] is a proper subinterval of [0. 1], Sharpening by Precision requires using the information from E such that $EP(H \mid E \wedge K) = [0.5, 0.5]$. More generally, any evidence that can be combined with K to derive a relatively precise reference class statement that is a candidate for the probability of H will increase the weight, because the interval from the new reference class statement will not conflict with the [0, 1] interval and it will be more precise. In any such scenario, the rule of Sharpening by Precision will require that one uses the more precise interval. Furthermore, this rule was developed independently of the problem of inertia, and therefore it is not just an *ad hoc* response to this issue.[4]

Levi criticises Kyburg's avoidance of inertia as *creatio ex nihilo*. His worry seems to be that Evidential Probability makes it possible to move from a state of complete ignorance to a precise probability distribution for H, without conditionalizing on a prior distribution. For the measurement of weight, the problem would be that Kyburg avoids inertia only by supposing evidence that is not present. However, regardless of whether Kyburg's overall system is satisfactory, his method of avoiding inertia is grounded in evidence: *if* we have information about a relative frequency that is more precise than the [0, 1] interval and *if* this is a suitable basis for the probability of a hypothesis H, then according to Kyburg this more precise relative frequency should be used. The Evidential Probabilist cannot shift from a [0, 1] interval for a hypothesis without suitable evidence that provides pertinent relative frequency data. One can approve or disapprove of Kyburg's method of creation, but it is incorrect to say that it proceeds from nothing, because the relative frequency data is essential.

Section Summary. In this section, I have reviewed some cases in which the WK measure works well, as well as some potentially difficult scenarios. The measure's successes in these areas does not prove that it is universally applicable. Indeed, it is plausible that there will be some forms of argumentation (perhaps mathematical argumentation) that are not amenable to analysis using Evidential Probability. Conceivably, the WK measure will be unable to measure weight in these areas. Nonetheless, the measure is promising and it seems broadly applicable. For example, unlike Kyburg in his 1968 article, I have not restricted the scope of an Evidential

---

[4]It might be *ad hoc* in other senses, but it was not specifically formulated to avoid the issue that Walley raises.

Probability-based measure of weight to arguments featuring evidence "of the same sort". However, one might wonder whether such the WK measure offers any worthwhile fruits. I shall now turn to an instance in formal epistemology where the WK measure can aide philosophical analyses.

3   The Paradox of Ideal Evidence

Popper's Paradox of Ideal Evidence (PIE), in [33] is a criticism of what he calls "subjective" theories of probability. By "subjective", Popper means theories that interpret probability (at least in some cases) as an epistemic concept concerning rational belief, rather than as a mind-independent concept like relative frequencies or propensities. Thus, theories like Subjective Bayesianism *and* Objective Bayesianism are "subjective" in Popper's sense. (He mentions the probability theories of Keynes and Carnap in particular.) Popper develops the following scenario: suppose that you have a coin, Z, when you have no knowledge of the bias or fairness of Z. Let H be the conjecture that "the nth unobserved toss of Z will be heads". Your background knowledge has no empirical evidence regarding H and you assign a probability of 0.5 to this conjecture. You subsequently learn E, which is a statistical report that is "ideally favourable" to this assignment of 0.5, such as a report stating that 1,500 out of the 3,000 tosses landed heads.[5] In your probability model for the coin, the tosses are not independent, such that earlier tosses of the coin can be relevant to subsequent tosses. Assume that the conditional probability of H given your new total evidence is equal to the prior probability, such that $P(H \mid E \wedge K) =$ 0.5. In some "subjective" probability theories, these values for the prior probability and the conditional probability will be mandatory, whereas in other theories (such as Subjective Bayesianism) they are merely permissible. In either case, Popper's example can be formulated.

There are actually two different problems that Popper develops from this scenario. I shall call these the 'Relevance Paradox' and the 'Representation Paradox'. Both are objections to "subjective" theories of probability, including Evidential Probability. I shall discuss them separately, and argue that the WK measure can help facilitate an answer to both paradoxes.

3.1   The Relevance Paradox

One of the putative applications of "subjective" theories of probability is that they can be used to define evidential relevance. The standard definition dates back to

---

[5]Popper's argument can also be made in terms of margins of error, but I shall simplify matters by focusing on the case where the sample frequency exactly matches the prior probability.

Keynes in [2]. Branden Fitelson notes that it has been popular among probabilists (including Bayesians) ever since Keynes's analysis [35]. On this definition, for some suitable "subjective" *precise* probability function P, any statement Φ, any hypothesis Ψ, and background knowledge Ω:

Evidential Relevance (1) Φ is evidentially relevant to Ψ, relative to Ω, if and only if $P(\Psi \mid \Phi \wedge \Omega) \neq P(\Psi \mid \Omega)$.

Put another way, in this analysis of evidential relevance, Φ is *evidentially* relevant to Ψ (relative to Ω) if and only if Φ is *probabilistically* relevant to Ψ (relative to Ω), for some suitable precise probability function P.

Popper's scenario raises a problem for this definition. The report of 3,000 tosses seems to be relevant to H, relative to the implicit background knowledge. However, since $P(H \mid E \wedge K) = P(H \mid K)$, the standard definition implies that E is not relevant to H in Popper's example. Thus, the first sub-paradox within Popper's PIE, which I have called "the Relevance Paradox", challenges analyses of evidential relevance that are based on "subjective" probability functions.

I shall not engage in a full critical discussion of the various responses that have been made to the PIE, since my objective is to use the WK measure to argue that there is *at least one* adequate answer to the Relevance Paradox, rather than that there is *only one* adequate answer. Nonetheless, I shall mention that Keynes seems to anticipate problems with his initial definition of relevance, and develops a "stricter and more complicated definition" on page 55 of [2] to address such problems. Unfortunately, Carnap, in section of [20], proved a trivialization problem for Keynes's strict definition: almost any arbitrary statement will be evidentially relevant to almost any other arbitrary statement. If a statement A that has an implication E that is probalistically relevant to H, then A is relevant to H on Keynes's "stricter" definition. Yet Carnap notes that, in classical logic, any statement A implies $(A \vee H)$, and this disjunction will be probabilistically relevant to H probability functions, except in special cases. (I explain the details of these special cases when I return to Carnap's trivialization problem.) Thus, one obvious adequacy condition for any answer to the Relevance Paradox is that it also addresses the issues that Carnap raises. I shall develop an answer that can address both challenges.

Consider what happens if we alter the standard definition to one using Evidential Probability:

Evidential Relevance (2) Φ is evidentially relevant to Ψ, relative to Ω, if and only if $\mathrm{EP}(\Psi \mid \Phi \wedge \Omega) \neq \mathrm{EP}(\Psi \mid \Omega)$.

(The only difference from Evidential Relevance (1) is that we have switched from a subjective precise probability function to the Evidential Probability function EP.)

Unlike precise probability functions, evidential probabilities can differ in two ways. Firstly, the means of the limits of the intervals can differ. For example, if $EP(H \mid E_1 \wedge K) = [0.8, 0.85]$ and $EP(H \mid E_2 \wedge K) = [0.5, 0.55]$, then the mean of the limits of the first probability is 0.825, whereas the value for the second is 0.525. Informally, the value of $EP(H \mid E_1 \wedge K)$ is "greater" than the value of $EP(H \mid E_2 \wedge K)$. Secondly, the degree of imprecision of the intervals can differ. For example, if $EP(H \mid E_1 \wedge K) = [0.1, 0.3]$ and $EP(H \mid E_2 \wedge K) = [0, 0.4]$, then the latter value is more imprecise, even though one cannot say that either probability value is "greater". Of course, it is also possible two intervals to contrast in both respects, such as $[0.1, 0.5]$ and $[0.9, 0.95]$. Therefore, if $EP(\Psi \mid \Phi \wedge \Omega) \neq EP(\Psi \mid \Omega)$, the difference might be a difference in mean value of the intervals, a difference in the precision of the intervals, or both.

The WK measure will increase or decrease when an increase or decrease in precision occurs, because the WK measure is proportionate to the degree of imprecision of evidential probabilities. We can use the WK measure to interpret how differences of precision can constitute evidential relevance: the additional information has either increased or decreased the weight of an argument from one's total evidence to the hypothesis in question. This enables an answer to the PIE that is intuitive and informally grounded, yet also systematic and formal: in Popper's scenario, the weight has increased. I shall now sketch this answer in more detail.

Suppose that you are almost totally ignorant about the coin in question. (Perhaps it has an extremely unusual shape, like a gömböc, that could result in a long-run frequency of 'heads' that is very high or very low.) In particular, suppose that you have no relative frequency knowledge that can be used to determine an evidential probability value with any precision. Therefore, the probability of the hypothesis H, that the coin will land heads, given your background information K, is $EP(H \mid K) = [0, 1]$. Such a position of extreme ignorance seems to be the sort of situation that Popper had in mind[6].

By combining E with your background knowledge, you might be able to use to infer that the long-run relative frequency of the coin landing heads is $0.5 \pm \epsilon$, where ϵ is 3%. Whether this value is correct will depend on whether it is acceptable, given your background knowledge, that the sample of tosses is representative, in the sense of matching the long-run relative frequency within a margin of error of 3%. (Your background knowledge might give you reasons to doubt that the sample is representative of the long-run relative frequency.) Assuming that you can extrapolate from

---

[6]My discussion on this point would not differ significantly if we assume that your have stronger background information, such as the knowledge that coins with this shape land at a relative frequency somewhere in the interval [0.01, 0.99] and that you can use this information to determine that $EP(H \mid K) = [0.01, 0.99]$.

E to the long-run relative frequency, you have learned from $(E \wedge K)$ that:

R_1 The long-run relative frequency of heads for this coin is between 0.47 and 0.53.

Since the nth toss is a member of the reference class of long-run tosses of the coin, it follows that the set of reference class statements for H has expanded, because it now includes R_1 and this statement is one of the statements that must be considered when determining $EP(H \mid E \wedge K)$. If R_1 is the appropriate reference class statement for H according to Sharpening, then $EP(H \mid E \wedge K) = [0.47, 0.53]$. Therefore, E can be relevant to H, relative to K, according to **Evidential Relevance (2)**. Using the WK measure, we can describe what has happened in Popper's example: E is relevant, but its relevance consists in increasing the weight.

One important aspect of this definition of evidential relevance and its interpretation using the WK measure is that evidence can be relevant by *decreasing* the weight of argument, as well as by increasing it. For example, in Seidenfeld's Hollow Cube scenario from Subsection 2.4, the second measurement is relevant to the hypothesis, but its relevance consists in decreasing the weight, as well as disconfirming the hypothesis that the volume is the cube is the high initial interval from the first measurement and the background knowledge.

As I mentioned earlier, Carnap proved that some definitions of evidential relevance suffer from a triviality problem. For example, on Keynes's strict definition, A is evidentially relevant to H, relative to K, if A classically entails E and E is probabilistically relevant to H given K. Carnap proved that, if $P(H \mid K)$ is neither 0 nor 1 and $P(A \mid K) \neq 1$, then A is evidentially relevant to H on Keynes's strict definition, for *any* statement A. Thus, 'There is a planet in the Solar System beyond Pluto' is relevant to 'Dark energy exists', because the former implies 'There is a planet in the Solar System beyond Pluto or dark energy exists', the this disjunction is probabilistically relevant to 'Dark energy exists', and neither statement has a value of 0 or 1 given current scientific knowledge.

In Evidential Probability, this trivialization proof will not work, because evidential probabilities can only differ if there is relative frequency data that alters that intervals via the Rules of Sharpening. Thus, if $(E \wedge K)$ does not imply any new reference class statements, then $EP(H \mid E \wedge K)$ will not differ from $EP(E \mid K)$, and E will not be relevant according to **Evidential Relevance (2)**. There are further issues for analyses of evidence that I have not addressed. For example, while the PIE suggests that probabilistic definitions of evidential relevance might be too narrow, there are some paradoxes in confirmation theory in which definitions of evidence are *arguably* too broad. These include the New Riddle of Induction (developed by Nelson Goodman in [37] and [38]) and the Paradox of the Ravens (developed by

Carl Hempel [39]). Neither paradox was originally targeted at probabilistic definitions of evidence, but they are further potential problems for any theory of evidence. Naturally I have not addressed these challenges and others in this paper. However, my argument does evince how one particular problem regarding evidential relevance can be addressed using the WK measure. In particular, the WK measure enables us to formalise how evidence can be relevant to a hypothesis *even if* that evidence neither confirms not disconfirms the hypothesis.

3.2 The Representation Paradox

One issue that Popper raises in light of this scenario is that, on standard "subjective" theories, an agent's degree of belief in H should not change upon learning E, and he claims that this entails that "subjective" probabilities cannot represent the change in weight, and therefore "subjective" probabilities cannot adequately represent evidential relations. However, the representation of evidential relations was part of the original motivation for the development of "subjective" probability theories. In contemporary terminology, Popper is claiming that probabilities cannot distinguish *ignorance*, the absence of relevant evidence, from *equivocation*, the presence of evenly-balanced evidence. This charge against theories like Bayesianism is still common today: see [40] and [41]. I shall call this issue the "Representation Paradox."

There are a number of possible lines of response for a "subjective" probabilist, but the WK offers a simple and direct reply to Popper: since weight is formalizable in terms of the imprecision of Evidential Probability intervals, it *is* possible to represent an increase in weight using "subjective" probabilities. The Evidential Probability intervals will narrow as the weight of argument for H given the total evidence increases. Thus, in Popper's scenario, prior to acquiring the report of tosses of the coin, you are in a state of complete ignorance with respect to the nth coin toss; once you have acquired the evidence from the report (and assuming that it can be combined with your background knowledge to enable you to utilise relatively precise frequency data about the coin) your total evidence leaves you in an equivocal position regarding the nth toss.

This shift from ignorance towards equivocation could have practical significance. Suppose that you were trying to estimate a precise expected value for H. (The context of the estimate might be gambling, but Popper's PIE is adaptable to a variety of possible situations.) In Chapter 14 of [17], Kyburg argues that the range of rational expected values is equal to the width of the Evidential Probability intervals. Thus, prior to learning any evidence about the coin, your choice of expected value is very arbitrary. However, if the interval for H given your total evidence becomes narrower, then only a value within this relatively narrow interval will be rational in Kyburg's decision theory. There is much more to be said here, as Kyburg's decision theory is fairly inchoate. Nevertheless, for the Representation Problem, the salient point is that if we assume that Evidential Probability is an adequate "subjective" probability system and that the WK measure is an adequate measure of weight, then "subjective" probabilities can represent what is happening in the PIE, and there will be *some* connection between the greater quantity of evidence and practical decisions.

Answering the Representation Paradox by appealing to the weight of argument is not an original move (for instance, see [7] and [42]) but the particular use of the WK measure is novel. Furthermore, there are some ancillary features of Kyburg's system which might make the WK particularly attractive as a means of formulating this response to Popper. For example, the width of Evidential Probability intervals are objectively determined, in the sense that they are always the same for given evidence. (See Chapter 12 of [16] for a detailed exposition of this point.) Put another way, two agents using Evidential Probability who have identical relevant evidence and a shared domain (the statements of a language) must have identical intervals for a given hypothesis, regardless of their subjective opinions. Since the WK measure is a function whose only variables are the limits of Evidential Probability intervals, this form of objectivity is also a feature of the WK measure. Consequently, on the WK measure, weight will be significantly independent of opinion. (There can still be a subjective element in what evidence we accept or what domain we choose.) Therefore, the WK measure can play a useful and novel role in answer to the Representation Paradox.

One possible objection to my answer is that a person might know that the precise relative frequency of 'heads' for tosses of a type of coin. In particular, they might know that the statistical statement that '50% of tosses of this type of coin land heads'. Suppose that this knowledge is the person's best reference class data, such that $EP(H \mid K) = [0.5, 0.5]$. No subsequent evidence for the frequency of 'heads' reference class will result in a more precise probability for the hypothesis given the total evidence. (This subsequent evidence could take the form of sample data for the type of coin, confirmation of a physical model for shapes with the coin's symmetries and asymmetries, analogical evidence from similar types of coins, and so on.) Yet their epistemic state does seem to have changed in some respect when such evidence is acquired.

My response to this objection is that (1) probabilities given evidence and (2) weight are collectively insufficient to characterise *all* the important aspects of someone's epistemic state, *but* that Evidential Probability can characterise the other aspects as well. In particular, the "subjectivist" can say that additional evidence can increase the reliability of one's existing evidence. This reliability can also be explicated using Evidential Probability. I mentioned Kyburg's formalisation of this concept at the end of Section 2.5: Kyburg develops a probabilistic model of the reliability of evidence as part of a broader model of scientific knowledge, in which evidence is ranked depending on its reliability given one's foundational evidence. Alternative probabilistic analyses of this aspect of reasoning include Richard Jeffrey's version of conditionalization [43] (in which we can model the evidence as merely probable) and the modelling of reliability by Luc Bovens and Stephan Hartmann in Chapter 3 of [44]. In the coin-tossing case, a probabilist can argue that it is the statistical statement that '50% of tosses of this type of coin' land heads that becomes more probable.[7] By using formalisations of relative probabilities *and* weight *and* reliability, a "subjectivist" can represent the evidential relations in such variations on Popper's PIE.

To formulate these answers to the Relevance Paradox and the Representation Paradox, I have made strong assertions and assumptions about the adequacy of both (a) Evidential Probability and (b) the WK measure. Since Evidential Probability currently has a very marginal position within imprecise probability theory (let alone the philosophy of science in general) I expect that both (a) and (b) will be objectionable to most readers. As my aim in this section has been to exemplify how the WK measure is philosophically useful, the fact that its adequacy plays an important part in my arguments above supports that aim, rather than undermines it. I have also not proven that these answers are superior to alternative responses to the PIE. However, my objective has been to develop a novel answer to the PIE, rather than to prove that this answer is uniquely adequate; it is plausible that there are multiple viable responses that defenders of "subjective" theories might make to Popper's paradox.

4   Conclusion

Weight seems to be an important part of argument strength. Keynes was sceptical about its quantitative measurement, but I have argued that the WK measure offers a propitious formalization of weight. Kyburg thought (at least from 1968 onwards) that measures of weight based on Evidential Probability would be limited to evidence "of the same sort", but once one drops the assumption that weight must increase monotonically, his approach can be expanded to a much more general measure. I have also illustrated how the WK measure can be used to formulate new arguments

---

[7]There could be further responses, such as denying that precise statistical generalisations can be acceptable, but I have accepted the details of the objection so that my defence is independent of such controversies.

within formal epistemology. I have not discussed alternative measures, such as Walley's measure in [24]. It might be the case that different measures of weight are preferable for analysing different arguments and inferences, so that alternative measures could be attractive even for ardent Evidential Probabilists. However, my aim has been to promote the WK measure, rather than to claim that other measures are unsatisfactory. I hope that my discussion has furthered this aim and illustrated the potential of Kyburg's system for the analysis of evidential relations.

Acknowledgements. I am grateful to Julian Reiss, Nancy Cartwright, Wendy Parker, Rune Nyrup, and the rest of the team at CHESS, for their assistance in the development of this article. I was also helped by a very encouraging and insightful group of referees.

References

[1] C. S. Peirce. The Probability of Induction. In C. Hartshorne and P Weiss, editors, *Elements of Logic*, volume 2 of *Collected Papers of Charles Sanders Peirce*, pages 82–105, Harvard, Mass. 1932. Harvard University Press.

[2] J. M. Keynes. *A Treatise on Probability*. Macmillan, London,

[3] D. A. Nance. The Burdens of Proof: Discriminatory Power, Weight of Evidence, and Tenacity of Belief. Cambridge University Press, Cambridge, 2016.

[4] R. M. O'Donnell. *Keynes: Philosophy, Economics and Politics*. Macmillan, Press, Basingstoke, 1989.

[5] I. J. Good. Weight of Evidence: A Brief Survey. *Bayesian Statistics*, 2: 249–270, 1985.

[6] J. Runde. Keynesian Uncertainty and the Weight of Arguments. *Economics and Philosophy*, 6 (2): 275–292, 1990.

[7] J. M. Joyce. How Probabilities Reflect Evidence. *Philosophical Perspective* 19 (1): 153-178, 2005.

[8] B. Weatherson. Keynes, uncertainty and interest rates. *Cambridge Journal of Economics*, 26 (1): 47-62, 2002.

[9] J. Hosiasson. Why Do we Prefer Probabilities Relative to Many Data? *Mind*, 40 (157): 23-26, 1931.

[10] B. Weatherson. The Bayesian and the Dogmatist. *Proceedings of the Aristotelian Society*, 107 (1pt2): 169–185, 2007.

[11] B. Davidson and R. Pargetter. Guilt Beyond Reasonable Doubt. *Australasian Journal of Philosophy*, 65 (2): 182–187, 1987. H.

[12] J. Franklin. Resurrecting Logical Probability. *Erkenntnis*, 55 (2): 277–305, 2001.

[13] J. Franklin. Case comment – United States vs. Copeland, 369 Supp. 2d 275 (E.D.N.Y. 2005): quantification of the 'proof beyond reasonable doubt' standard. *Law, Probability and Risk*, 5 (2): 159–165, 2006.

[14] Kyburg. *Probability and the Logic of Rational Belief*. Wesleyan University Press, Middletown Connecticut, 1961.

[15] H. E. Kyburg. Bets and Beliefs. *American Philosophical Quarterly*, 5 (1): 54–63, 1968.

[16] H. E. Kyburg and C. M. Teng. *Uncertain Inference*. Cambridge University Press, Cambridge, 2001.

[17] H. Reichenbach. *The Theory of Probability*. Berkeley, University of California Press, 1949.

[18] H. E. Kyburg. *The Logical Foundations of Statistical Inference*. Dordrecht, Reidel, 1974.

[19] H. E. Kyburg. *Science and Reason*. Oxford University Press, New York, 1990.

[20] R. Carnap. *The Logical Foundations of Probability*. University of Chicago Press, Chicago, 1962.

[21] *Synthese*, 186 (2), 2012.

[22] R. White. Evidential Symmetry and Mushy Credence. In T. Szabó Gendler and J. Hawthorne, editors, *Oxford Studies in Epistemology*, pages 161–186, Oxford. 2010. Clarendon Press.

[23] A. P. Pedersen and G. Wheeler. Demystifying Dilation. *Erkenntnis*, 79 (6): 1305–1342, 2014.

[24] Bradley, Seamus, "Imprecise Probabilities", *The Stanford Encyclopedia of Philosophy* (Summer 2015 Edition), Edward N. Zalta (ed.), http://plato.stanford.edu/archives/sum2015/entries/imprecise-probabilities/

[25] T. Seidenfeld. Forbidden Fruit: When Epistemological Probability may *not* take a bite of the Bayesian apple. In W. Harper and G. Wheeler, editors, *Probability and Inference: Essays in Honour of Henry E. Kyburg, Jr.*, pages 267–279, London. 2007. College Publications.

[26] H. E. Kyburg. Bayesian Inference with Evidential Probability. In W. Harper and G. Wheeler, editors, *Probability and Inference: Essays in Honour of Henry E. Kyburg, Jr.*, pages 281–296, London. 2007. College Publications.

[27] P. Walley. *Statistical Reasoning with Imprecise Probabilities*. Chapman, London, 1991.

[28] I. Levi. Probability Logic and Logical Probability. In W. Harper and G. Wheeler, editors, *Probability and Inference: Essays in Honour of Henry E. Kyburg, Jr.*, pages 255–261, London. 2007. College Publications.

[29] K. Popper. *The Logic of Scientific Discovery*. Hutchinson, London. 1980.

[30] B. Fitelson. Goodman's "New Riddle". *Journal of Philosophical Logic*, 37 (6): 613-643, 2008.

[31] N. Goodman. A Query on Confirmation. *The Journal of Philosophy* 43 (14): 383-385, 1946.

[32] N. Goodman. *Fact, Fiction, and Forecast*. The Athlone Press, University of London,

1954.

[33] C. G. Hempel. Studies in the Logic of Confirmation (I.). *Mind*, 54 (213): 1-26, 1945.

[34] J. D. Norton. Challenges to Bayesian Confirmation Theory. In S. Bandyopadhyay and M. R. Forster, editors, *Philosophy of Statistics*, pages 391–440, Oxford. 2011. Elsevier B. V.

[35] J. Reiss. What's Wrong With Our Theories of Evidence? *Theoria: An International Journal for Theory, History and Foundations of Science* 29 (2): 283-306, 2014.

[36] R. O'Donnell. Keynes's Weight of Argument and Popper's Paradox of Ideal Evidence. *Philosophy of Science*, 59 (1): 44-52, 1992.

[37] R. C. Jeffrey. *The Logic of Decision*. New York: McGraw-Hill, 1965.

[38] L. Bovens and S. Hartmann. *Bayesian Epistemology*. Oxford: Clarendon Press, 2003.

Received 14 May 2017

On Elitist Lifting and Consistency in Structured Argumentation

Sjur Dyrkolbotn

Department of Civil Engineering, Western Norway University of Applied Sciences
sdy@hvl.no

Truls Pedersen

Department of Information Science and Media Studies, University of Bergen

Jan Broersen

Department of Philosophy and Religious Studies, Utrecht University

Abstract

We address the question of how to lift an ordering over rules to an ordering over arguments (sets of rules) that is well-behaved. It has been shown that so-called elitist lifting may lead to inconsistencies. We give restrictions on the underlying rule-ordering that avoid inconsistency. Then we show that a recently proposed solution – so-called disjoint strict lifting – that was introduced to address conceptual objections, also leads to inconsistency. We show that another recent proposal, telling us to reorder rule-orderings to take argument structure into account before lifting, is able to avoid conceptual problems without leading to any new inconsistencies. We generalise this approach by defining what we call structural rule-orderings and show a correspondence between weakest link and last link lifting of such orderings, which has interesting consequences for the question of consistency. We arrive at our results using a signature-based approach to structured argumentation. Instead of settling on a given framework, such as ASPIC$^+$, we define an argumentation language that allows us to express only those properties of argumentation systems we need to establish our results. This abstract approach simplifies and clarifies the technical work while making our contribution more general.

Keywords: Argumentation, Preferences, Weakest Link, Elitist Lifting, Consistency, Rationality.

Dyrkolbotn and Broersen gratefully acknowledge financial support from the ERC-2013-CoG Project REINS, No. 616512.

1 Introduction

Structured argumentation is about combining defeasible and non-defeasible reasoning rules into arguments that can be studied in terms of their structure. Abstract argumentation is about what arguments count as valid arguments in the context of other arguments, independently of the internal structure of the arguments themselves. Any argumentation system based on this two-tiered view will have to ensure congruence between the semantics of the underlying defeasible reasoning systems whose rules are used to build arguments, and the abstract semantics that governs the interaction between arguments [3, 12, 11].

Typically, the semantics for abstract argumentation is provided by Dung's theory of argumentation frameworks [6], which looks only at the relations between different arguments, providing a semantics using graph-theoretic definitions. In Dung's theory, arguments have no internal structure whatsoever. Therefore, applying abstract argumentation to a concrete argumentation scheme can be looked at as a process of abstraction and transformation: Dung-style semantics transforms arguments characterised by their content into arguments characterised by their relationship with other arguments.

This raises a number of issues regarding the relationship between the semantics of the underlying defeasible reasoning system and the 'abstract' relational semantics. One of the most fundamental questions that arise is the question of consistency: if we translate arguments into vertices in a graph and apply Dung's semantics, do we get a theory extension that is consistent in the underlying logic? *Prima facie*, we seem entitled to expect so [3]. After all, arguments should be sound, and accepted arguments should be sound in the underlying logic. However, as recent research on structured argumentation has shown, ensuring consistency is far from trivial, especially when preferences are used to prioritise between different rules and arguments [11, 4, 15, 10, 9]. In this paper we study the consistency problem, focusing on the question of how to lift priority orderings over rules to preference orderings over arguments, without violating consistency.

The structure of this paper is as follows. In Section 2 we provide some motivation and discuss a challenging example that has motivated recent work. In Section 3 we define basic concepts and present a signature-based approach to structured argumentation, allowing us to be clear about the assumptions we need for proving results, without sacrificing generality. In Section 4, we introduce preferences into the mix and discuss the relationship between orderings over rules and preferences between arguments, an issue at the heart of the consistency problem. We then give an order-theoretic constraint that ensures consistency for a broad range of argumentation systems, generalising a similar constraint found in [11]. Following up on this, we give two results that link the order-theoretic constraint with the so-called elitist approach to lifting orderings over rules to preferences over arguments. In Section 5 we take a step back and consider a conceptual problem that has led other researchers to propose disjoint elitist lifting as an alternative approach to connecting rule-orderings and preferences [10, 15]. We show that the approach is flawed, since it cannot be generalised to systems that allow strict rules (with so-called "restricted rebut"). If such rules are allowed, inconsistencies arise even for linear orders. Specifically, it is shown that disjoint lifting is no answer to the consistency problem for ASPIC$^+$, a leading framework for structured argumentation. In Section 6 we consider another recent proposal from the literature, asking us to ensure that rule-orderings reflect the order in which rules can be applied when arguments are constructed (if necessary, we might have to reorder the inital ordering of rules to satisfy this constraint) [14, 15]. Unlike the proposal for disjoint lifting, this proposed shift of attention towards *structural* orderings is shown to have several appealing consequences. A summary of results and a conclusion is offered in Section 7.

2 Motivation

If all rules are defeasible or strict rule applications are vulnerable to objections (which would mean they are not actually "strict"), ensuring consistency is not a deep problem. In this case, when two arguments have contradictory conclusions, any reasonable translation scheme will ensure that at least one of them defeats the other in the corresponding abstract argumentation framework. It will follow from any reasonable abstract semantics that not both arguments can be accepted. Hence, the theory resulting from taking the conclusions of all accepted arguments in a given Dung extension will be consistent (assuming natural closure properties that ensure the existence of a corresponding argument for every logical consequence of a consistent theory).

However, if there are strict rules in the system, that is, rules whose application is beyond objection (we assume, of course, that all preconditions are met so the rules are active), ensuring consistency becomes non-trivial. For a concrete example of this phenomenon, consider a system with defeasible rules r_1, r_2 and strict rules s_1, s_2 and s_3, depicted below.

$$r_1: \Rightarrow p_1, \qquad r_2: \Rightarrow p_2,$$
$$s_1: \ p_1 \rightarrow p_3, \quad s_2: \ p_2 \rightarrow p_4, \quad s_3: \ p_4 \rightarrow \neg p_3$$

By applying these rules in the ordinary manner of logic, we can form the following arguments (proof trees), where a curved line indicates that a defeasible rules is being used to infer the conclusion (a strict rule application is indicated by a straight line, as usual).

$$A: \quad \dfrac{\dfrac{\overset{\sim\!\sim}{p_2}}{p_4}}{\neg p_3} \qquad B: \quad \dfrac{\overset{\sim\!\sim}{p_1}}{p_3}$$

A and B have contradictory conclusions, so to ensure consistency we must avoid accepting both of them. However, the defeasible rules in A and B are r_1 and r_2 respectively, which are not in any direct opposition to one another. Specifically, A and B will only attack each other if our system permits attacks on strict rule-applications. In this article we work with systems that do not allow this, described in the literature as systems with restricted rebuttal. For such systems, with respect to A and B above, we cannot exclude either argument, which means that our extension will be inconsistent.

In a system without preferences, the natural solution to this problem is to ensure that we can reason contrapositively with strict rules. If A and B have contradictory conclusions and A concludes with a sequence of strict rules, there should be some way of using B in combination with contrapositive reasoning with strict rules from A to obtain an argument B' that contradicts the conclusion of some defeasible rule applied earlier in A. In our example we need to ensure that at least one of $\bar{R}_1$ and $\bar{R}_2$ exists, where the overbar notation is used to indicate opposition (e.g., $\bar{R}_1$ denotes an argument in opposition to any argument that includes r_1):

$$\bar{R}_1: \quad \dfrac{\dfrac{\dfrac{\overset{\sim\!\sim}{p_2}}{p_4}}{\neg p_3}}{\neg p_1} \qquad \bar{R}_2: \quad \dfrac{\dfrac{\dfrac{\overset{\sim\!\sim}{p_1}}{p_3}}{\neg p_4}}{\neg p_2}$$

The final rule applied in $\bar{R}_1$ is a contrapositive application of s_1, while the final two rules applied in $\bar{R}_2$ are contrapositive applications of s_2 and s_3 respectively. Notice that $\bar{R}_1$ is an argument that uses only one defeasible rule, namely r_2. Symmetrically, $\bar{R}_2$ is an argument with r_1 as the only defeasible rule. Hence, we have captured the opposition between these two rules at the argument level, ensuring also that A and B will not both be accepted by any reasonable argumentation semantics (since they include r_2 and r_1 respectively).

It has been shown that if the argumentation system we use satisfies some natural properties (e.g., that B attacks A when it attacks one of the subarguments of A), contraposition will suffice to ensure consistency, even in the presence of strict rules [3,

12]. However, if the system takes preferences into account, the consistency problem becomes harder. In this case, it is natural to hold that $\bar{A}$ can only defeat A by contradicting the conclusion of A when A is not strictly preferred to $\bar{A}$ (this is how preferences are used to prune the attack relation in the ASPIC$^+$ framework). This puts a new obstacle in our path to consistency. For instance, if A is strictly preferred over $\bar{R}_2$ and B is stricly preferred over $\bar{R}_1$ in the example above, the attacks we need to avoid inconsistency will be rendered impotent by the preference relation.

What this means is that we need to put restrictions on our preference relations over arguments, to ensure that a preference-dependent translation from logic-based arguments to graph-based argumentation frameworks does not result in inconsistency. In the example above, it seems perfectly reasonable to demand either that $\bar{R}_2$ is no worse than A or that $\bar{R}_1$ is no worse than B, depending on the relationship between r_1 and r_2. Indeed, if some argument A contains no defeasible rule strictly worse than any defeasible rule in B, it is hard to see how A could be a strictly worse argument than B. For $\bar{R}_1$ and $\bar{R}_2$, this situation must obtain in one direction or the other, so the requirement we need seems fundamentally reasonable in this case.

When assessing whether a given preference ordering is reasonable in this context, we tend to look at the relationship between how we order arguments and how we order rules. Typically, the assumption is that the latter is an atomic parameter set in advance, e.g., in the form of a priority or specificity ordering that comes from the underlying logic. But nothing really hinges on this assumption. The broader point is that we have strong expectations that arguments built using defeasible rules need to be ordered in a way that allows us to explain the ordering in terms of a corresponding ordering over these rules.

In the following, we will assume that the relationship between preferences over arguments and priorities over rules is of the so-called *elitist* variety. Roughly, what this means is that A is less preferred than B whenever there is a rule among the defeasible rules in A that is less preferred than all the defeasible rules that occur in B. However, the elitism criterion may be directed to specific subsets of the defeasible rules in arguments. Depending on the subsets of defeasible rules we focus on, we get different varieties of elitism, three of which we will explore later in this article. If we limit the scope of the principle to cases where there is a relevant rule in A that is *strictly* less preferred than all relevant rules in B, we get a variety that we refer to as *strict elitism*.

Before turning to the formal details, we give a final motivating example to show that there is an interesting non-trivial tension between ensuring consistency and giving a reasonable treatment of preferences. The example we give is due to Dung [7], who asks us to consider a dilemma involving four rules, $r_1, \ldots, r_4$, each of which has the empty set as the only premise and an associated conclusion $c(r_i)$. This becomes a dilemma because we have strict rules available, which include s_l: $c(r_i), c(r_j), c(r_k) \rightarrow \neg c(r_l)$ for all bijections between $\{i, j, k, l\}$ and $\{1, 2, 3, 4\}$. That is, the negation of the conclusion of any one of the r-rules can be derived from the conclusions of the three other r-rules by a strict rule. In terms of the underlying logic, what this means is that taking the conclusions of all four r-rules yields an inconsistent theory.

Avoiding this outcome becomes difficult if we assume an ordering over rules that partitions the r-rules into two incomparable equivalence classes where $r_1 \equiv r_2$ and $r_3 \equiv r_4$ (so that all other rule-pairs are incomparable). This leaves us with a set of dilemmas, where we have to choose between arguments $\bar{A}_i, A_i$ of the following form:

$$\bar{A}_i: \quad \frac{\overbrace{c(r_j)}, \overbrace{c(r_k)}, \overbrace{c(r_l)}}{\neg c(r_i)}, \qquad A_i: \quad \overbrace{c(r_i)}$$

Clearly, $\bar{A}_i$ and A_i are in opposition for all $i \in \{1, 2, 3, 4\}$. Moreover, there is a *prima facie* reason to think that A_i is preferable to $\bar{A}_i$, in view of the priority ordering given over the rules. Granted, $\bar{A}_i$ concludes with a strict rule, but it uses three defeasible rules to get there, one of which is at least as bad as r_i. Hence, $\bar{A}_i$ as a whole seems at least as bad as A_i. Moreover, since $\bar{A}_i$ contains two defeasible rules not comparable to any rule used in r_i, there is an intuitive reason to think that A_i is *strictly better* than $\bar{A}_i$. Indeed, this intuition corresponds to the judgement prescribed by the version of elitist lifting used in the first version of ASPIC$^+$ [11] (see Definition 11 below for the formal details).

Unfortunately, if we accept this intuitive judgement, we arrive at an inconsistency. Indeed, if we reason as we did above for all A_i, then we end up ignoring all attacks made on these arguments by the corresponding $\bar{A}_i$ (here we assume that attacks are ignored when their source is strictly less preferred than their target, as is often the case in structured argumentation). But then we end up accepting all A_i, resulting in a set of arguments that yields an inconsistent theory.

When it was presented for the first time – in a slightly different context – the example above came as a surprise. Specifically, the authors of [11] realised that it revealed an error in a theorem regarding the leading framework of structured argumentation systems, ASPIC$^+$. The solution was to replace elitism by strict elitism, for which consistency can be ensured [11, corrected online 2017]. In the Dung example, this solves the problem by rendering all argument pairs in the dilemma incomparable, so that none is strictly better than any of the others (so the attacks needed to prevent inconsistency are not blocked). However, strict elitism is conceptually problematic for partial rule orderings, when some rules are incomparable. In this case, if A contains a least preferred rule r_1 and B is an argument that consists only of r_2, with $r_1 \equiv r_2$, A and B will be considered equivalent or incomparable regardless of what other rules A contain. However, if A contains additional defeasible rules that are not comparable to r_1 and r_2, we seem entitled to think that B is a strictly better argument (which is the judgement prescribed under non-strict elitism).

In view of this and other concerns, some of which are addressed in more depth in Section 5, there is still more work needed on the question of how to determine the relationship between orderings over rules and orderings over arguments, without violating consistency. In the following, we present our own results on this question, adopting an abstract approach that we believe is suited for delivering general results that are not tailor-made for specific argumentation systems, but focus on the properties such systems should meet in order to make sense of rule orderings and preferences.

3   A signature for structured argumentation

Unlike most other works in structured argumentation, we adopt a top-down approach, starting from a set $\mathcal{A}$ of atomic argument labels (henceforth referred to simply as arguments) and an attack relation $\mathsf{Att} \subseteq \mathcal{A} \times \mathcal{A}$. An argumentation system is then some structure A that instantiates $(\mathcal{A}, \mathsf{Att})$ by telling us how arguments are built, how the relation of attack is derived from their structure, and which (sets of) arguments we may accept. The most abstract example of an argumentation system (in this sense) is Dung's theory of argumentation frameworks, where both $\mathcal{A}$ and Att are taken as primitives, yielding a directed graph with a purely graph-theoretical semantics [6]. Another important class of argumentation systems arise from Prakken's ASPIC$^+$ framework, where arguments are proof trees in default logic whose internal structure determine the relation of attack, a relation that is subsequently refined using preferences, before Dung's theory is used to provide a semantics [12, 11].

In this article, we will not limit our investigation to any specific argumentation system. Rather, we will characterise properties of such systems and prove results about them in terms of the following language.

Definition 1 (Argumentation Language). An *argumentation language* is a structure $\mathfrak{X} = (\mathcal{L}, \mathbb{S}, \mathbb{D}, \leq, \mathcal{A}, \mathsf{Att})$ which provides:

- A logical language $\mathcal{L}$ that comes also with a contrariness function $^{-} : \mathcal{L} \to 2^{\mathcal{L}}$.

- A set of strict rules $\mathbb{S}$.

- A set of defeasible rules $\mathbb{D}$ and a *reflexive and transitive* ordering $\leq \subseteq \mathbb{D} \times \mathbb{D}$.

- A carrier set $\mathcal{A}$ of all arguments and an attack relation $\mathsf{Att} \subseteq \mathcal{A} \times \mathcal{A}$.

- For each argument $A \in \mathcal{A}$ we also have:
  - a conclusion: $c(A) \in \mathcal{L}$,
  - a set of defeasible rules which occur in an argument: $\mathcal{R}_d(A) \subseteq \mathbb{D}$, and
  - a set of subarguments $Sub(A) \subseteq \mathcal{A}$.

All the concepts above are instantiated by any argumentation system obtained using the ASPIC$^+$ framework. However, many other systems can also be viewed as instantiations of argumentation languages under the perspective adopted here. Indeed, even an abstract argumentation framework – a directed graph – can be viewed as an instantiation of an argumentation language. The trivial such instantiation takes $\mathcal{A} = \mathbb{D} = \mathcal{L}$ as the set of arguments, lets Att be the relation of attack, leaves $\leq$, $\mathbb{S}$ and $^-$ empty, and maps the remaining components to the identity function. Of course, this instantiation is not particularly interesting. However, it illustrates that there are many ways of associating a given argumentation system with an argumentation language, some more interesting than others. For systems obtained using ASPIC$^+$, the (intended) corresponding argumentation language is obvious, since the constructs above are constructively defined in such systems (as required by the ASPIC$^+$ template). For other systems, such as sequent-based [1] or assumption-based argumentation [2], it is less obvious how we should instantiate the corresponding argumentation language. It can be done in a number of different ways, but the question is how we ought to do it to establish interesting results. We will not investigate this question here, for any non-ASPIC$^+$ argumentation system. We note, however, that *if* it is possible to establish a mapping such that the conditions of our results hold, then those results apply to non-ASPIC$^+$ systems as well. This, we believe, is an added benefit of a signature-based approach to structured argumentation, which makes the results generalisable through appropriate instantiations.

As usual, the contrariness function $^- : \mathcal{L} \to 2^{\mathcal{L}}$ collects those formulas that are *contrary* to a given formula. How this set is defined depends on the underlying logic. We assume that if $\Gamma \subseteq \mathcal{L}$ is a theory of this logic and there is some $\phi \in \Gamma$ such that $\Gamma \cap \overline{\phi} \neq \emptyset$, then Γ is an inconsistent theory. This is all we need to know about the underlying logic in the present article. For simplicity, we also assume that $\phi \in \overline{\psi}$ if, and only if, $\psi \in \overline{\phi}$ for all $\phi, \psi \in \mathcal{L}$. That is, contrariness is symmetric. Nothing much hinges on this assumption, but it will simplify the statement of some definitions and results. It is also reasonable, e.g., we expect $\phi \in \overline{\neg\phi}$ whenever $\neg\phi \in \overline{\phi}$.

Note that the ordering $\leq$ over the rules is assumed to be reflexive and transitive. This is a standard requirement for any ordering that encodes a notion of strength or preference, which is conceptually and independently motivated. In the following, we will consider what additional constraints on $\leq$ we need in order to ensure consistency when the ordering over the rules is lifted to an ordering over arguments.

The elements that make up an argumentation language can be refined in various ways, to characterise different argumentation systems over the same language. In the following definition, we record the signature of some of the most important refinements that we need in order to characterise existing systems, including ASPIC$^+$.

Definition 2 (Argumentation Signature). Given an argumentation language $\mathfrak{X}$, an (argumentation) signature is a tuple $\Omega = (\mathsf{LastRule}, \mathcal{R}_d^S, F_S, M_S, \mathsf{Def})$ where for every argument $A \in \mathcal{A}$:

- $\mathsf{LastRule}(A) \in \mathbb{S} \cup \mathbb{D}$ is the last rule applied in A.

    - Intuitively, this encodes the idea that arguments have a proof-theoretic structure; they conclude with a unique rule that establishes the conclusion of the argument as a whole.[1]

- $\mathcal{R}_d^S(A) \subseteq \mathcal{R}_d(A)$ is the set of *salient* defeasible rules that occur in A.

    - Intuitively, a defeasible rule is salient in A if we may take it into account when comparing A with some other argument to determine the relative preference of A.

- $F_S(A) \subseteq Sub(A)$ is the set of *fallible subarguments* of A.

    - Intuitively, the fallible subarguments of A are those subarguments of A that are open to direct attack (as opposed to an implicit attack via a fallible subargument).

- $M_S(A) \subseteq F_S(A)$ is the set of *maximally fallible* subarguments of A.

    - Intuitively, the maximally fallible subarguments are the fallible subarguments of A that are maximal in some sense, e.g., because they are not themselves subarguments of any fallible subargument of A.

- $\mathsf{Def}(A) \subseteq \mathsf{Att}(A)$ is the set of arguments defeated by A.

---

[1]The last rule of an argument is assumed given as a primitive, it is not constructively defined, so we only define its type. The reason for this is that we want to ensure that our work remains general; the results we prove apply to all systems satisfying the conditions of our results, regardless of how exactly the function $\mathsf{LastRule}$ is instantiated in a given context. We follow this definition strategy for all the components of an argumentation signature; indeed, this is why it is referred to as a signature.

– Intuitively, the arguments defeated by A are the arguments attacked by A after pruning the attack relation to take preferences into account.

To illustrate how we work at the signature level, consider the notion of a fallible subargument. All we have said is that such a notion belongs to an argumentation signature. We have not said anything about how it should be defined, beyond providing the underlying intuition. However, we can offer a (partial) formal characterisation of fallible subarguments by stipulating the following constraint, for all $A \in \mathcal{A}$:

$$F_S(A) = \{B \in Sub(A) \mid \mathsf{LastRule}(B) \in \mathbb{D}\} \tag{1}$$

This constraint is satisfied in ASPIC$^+$. However, it is not satisfied by all argumentation systems. Both the system in [4] and the systems in [10] are better characterised by $F_S(A) = Sub(A)$ for all $A \in \mathcal{A}$. We do not have to exclude systems that violate our constraint; the signature is purely descriptive and should not be given a normative reading. Moreover, there is no reason to exclude from consideration a system that fails to include its own definitions of what it means to be a fallible or a maximally fallible subargument. For such systems, the task is to analyse their behaviour and *infer* what notion of fallible and maximally fallible subargument they use.

Recall that we defined a possible refinement of defeasible rules into salient defeasible rules. This allows us to characterise the difference between weakest link and last link reasoning, in the following sense:

Weakest Link Salience: $\mathcal{R}_d^S(A) = \mathcal{R}_d(A)$, for all $A \in \mathcal{A}$, meaning that the least preferred defeasible rule is salient, i.e., the weakest link is salient.

Last Link Salience: $\mathcal{R}_d^S(A) = \{r \in \mathbb{D} \mid \exists B \in M_S(A) : \mathsf{LastRule}(B) = r\}$, for all $A \in \mathcal{A}$, meaning that the least preferred defeasible rule is only salient if it is the final rule of a maximally fallible subargument, i.e., the weakest link is only salient if it is a last link.

Similarly, we can characterise a key element of preference-independent argumentation systems (including Dung's theory) by stipulating that there is no distinction between attack and defeat, i.e., $\mathsf{Att}(A) = \mathsf{Def}(A)$ for all $A \in \mathcal{A}$.

The first constraint we will assume to hold quite generally, for all argumentation systems A, is the following, ensuring that the conclusion of an argument is the conclusion of the last rule applied in that argument.

$$\forall A \in \mathcal{A} : r = \mathsf{LastRule}(A) \Rightarrow c(A) = c(r) \tag{2}$$

Secondly, we impose the following constraint, encoding the intuition that maximally fallible subarguments are fallible subarguments that are not subarguments of any other fallible subargument.

$$M_S(A) = \{B \in F_S(A) \mid \neg \exists C \in F_S(A) \text{ s.t. } Sub(B) \subset Sub(C)\} \qquad (3)$$

This constraint is satisfied both by the systems found in [4, 10] (where $M_S(A) = \{A\}$ for all $A \in \mathcal{A}$) and by ASPIC$^+$, showing a point of agreement between distinct systems.

Thirdly, we impose the following constraint, encoding the intuition that defeats target fallible subarguments.

$$\forall A, B \in \mathcal{A} : B \in \mathsf{Def}(A) \Leftrightarrow \exists B' \in F_S(B) : B' \in \mathsf{Def}(A) \qquad (4)$$

That is, A defeats B if, and only if, A defeats some fallible subargument of B. In addition, we make the following assumption about arguments that are defeated without being defeated on any strict subargument. It encodes the intuition that defeats effectively emanate from attacks on rules.

$$\forall A, B \in \mathcal{A} : (A \in \mathsf{Def}(B) \text{ and } \forall C \in F_S(A) \setminus \{A\} : C \notin \mathsf{Def}(B)) \Rightarrow \\ (\forall D \in \mathcal{A} : \mathsf{LastRule}(D) = \mathsf{LastRule}(A) \Rightarrow D \in \mathsf{Att}(B)) \qquad (5)$$

This constraint effectively says that arguments can only be successfully attacked on their rules: if B defeats A without defeating any fallible proper subargument of A, then B attacks every argument concluding with the same rule as A (but the attack might not be a defeat, depending on how we prune the attack relation). We also make the following assumption about the attack relation, for all $A, B \in \mathcal{A}$:

$$c(B) \in \overline{c(A)} \text{ and } \mathsf{LastRule}(A) \in \mathbb{D} \Rightarrow A \in \mathsf{Att}(B) \qquad (6)$$

This constraint implies that the system includes a certain type of attack between arguments, referred to as restricted rebuttal in the theory of structured argumentation, see e.g., [12]. Notice how this kind of attack links the notion of contrariness from the underlying logic with the argumentation-theoretic concept of an attack between arguments. Restricted rebuttal is the only notion of attack we need to include explicitly in order to discuss the consistency problem addressed in this work. That is, our results require the presence of *at least* all attacks corresponding to rebuttals of defeasible rules. This does not exclude systems that include *additional* attacks, e.g., undercutting attacks of the type discussed in [12]. Our results still hold for such systems, as long as all the constraints we require are satsified. The point is that our results do not depend on the particularities of other types of attack, so the signature-based approach allows us to abstract away from them to deliver a more general treatment of the consistency problem.

Systems including so-called *unrestricted rebuts* deserve a special mention, however, since they give rise to additional challenges that we do not address in this work. These additional challenges arise from the fact that such systems allow attacks also on strict rule-applications, as discussed at length in [4, 9]. This means that certain desirable closure properties are not trivially satisfied for these systems, see also the discussion after Definition 5 below. The reader should note that while attacks targeting the conclusions of strict rules are not ruled out by (6), they are ruled out whenever conditions (1), (4) and (5) all hold. In this case, all attacks must target the last rule of a fallible subargument, all of which must in turn be defeasible.

The next property we assume to hold is transitivity of the fallible subargument relation: all fallible subarguments of B are also fallible subarguments of all arguments that have B as a fallible subargument.

$$\forall A, B \in \mathcal{A} : B \in F_S(A) \Rightarrow \forall C \in F_S(B) : C \in F_S(A) \tag{7}$$

This constraint is satisfied by ASPIC^+ and all other systems of argumentation of which we are aware. In addition, we will assume that the following property holds.

$$\forall A \in \mathcal{A} : \forall B \in M_S(A) : M_S(B) = \{B\} \tag{8}$$

Intuitively, this constraint says that if an argument is a maximally fallible subargument of some other argument, then it is also the only maximally fallible subargument of itself. Finally, we will be assuming the following, stating that any two strict arguments have mutually consistent conclusions.

$$\forall A, B \in \mathcal{A} : (\mathcal{R}_d(A) \cup \mathcal{R}_d(B) = \emptyset) \Rightarrow c(A) \notin \overline{c(B)} \tag{9}$$

Notice that this only requires that all *arguments* consisting of only strict rules have mutually consistent conclusions. It is not required that all strict rules are mutually consistent. If two strict rules are inconsistent, our constraint only implies that establishing their preconditions requires us to use at least one defeasible rule. Effectively, the restriction is a consistency requirement on the underlying logic: if two formulas are contradictory according to the contrariness relation of this logic, then the strict rules from this logic must not allow us to prove that both formulas are tautologies (notice that this does not rule out paraconsistent background logics, for which the contrariness function must be defined differently than in classical logic).

From now on, when we say that A is an *argumentation system*, we mean that it is a structure that instantiates all operators from Definitions 1 and 2, while satisfying

(2)-(9). We remark that all of these properties are independent and required for our proofs – each of them (or stronger properties that imply more than one of them) must therefore be established to conclude that our results apply to a non-ASPIC$^+$ system. In the next section, we give the signature of Dung-style argumentation semantics, before stating a consistency requirement that targets the relationship between the argumentation semantics and the semantics of the underlying logic.

3.1 Argumentation semantics and consistency

By now, we have encountered three distinct notions of opposition between arguments.

- First, we can have $c(A) \in \overline{c(B)}$, meaning that the conclusion of A is contrary to the conclusion of B according to the underlying logic.

- Second, we can have $B \in \mathsf{Att}(A)$, meaning that B is attacked by A.

- Third, we can have $B \in \mathsf{Def}(A)$, meaning that B is defeated by A.

The relation between the three notions of opposition identified above is very important in structured argumentation theory. It pertains to the consistency problem, as we will see shortly. First, let us define (Dung-style) semantics for argumentation systems.

Argumentation semantics obtained via Dung's theory of argumentation frameworks only look at the pair $(\mathcal{A}, \mathsf{Def})$ to determine which collections of arguments we may accept. This leads to the following signature, where we add two constraints that are satisfied by all main argumentation semantics.

Definition 3 (Abstract semantics). An argumentation semantics for A is an interpretation function ε such that $\varepsilon(\mathsf{A}) = \varepsilon(\mathcal{A}, \mathsf{Def}) \subseteq 2^{\mathcal{A}}$ for all $\mathcal{A}, \mathsf{Def}$, where the different collections of arguments included $\varepsilon(\mathcal{A}, \mathsf{Def})$ are referred to as *(abstract) extensions*. We require that ε satisfies the following constraints:

- Conflict-freeness: for all $E \in \varepsilon(\mathcal{A}, \mathsf{Def})$:

$$\forall A, B \in E : A \notin \mathsf{Def}(B)$$

  – Intuitively, there are no attacks between any two arguments in the same extension.

- Completeness: for all $E \in \varepsilon(\mathcal{A}, \mathsf{Def})$:

$$\forall A \in \mathcal{A} : ((\forall C \in \mathcal{A} : A \in \mathsf{Def}(C) \Rightarrow \exists D \in E : C \in \mathsf{Def}(D)) \Rightarrow A \in E))$$

    – Intuitively, if an argument is defended by an extension, then it must be included in that extension.

An argumentation semantics for $(\mathcal{A}, \mathsf{Def})$ provides a semantics for argumentation also at the level of $\mathcal{L}$, in the following sense.

Definition 4 (Concrete semantics). We lift the conclusion function to also apply to extensions. If A is an argumentation system, and $E \in \varepsilon(\mathsf{A})$ is an extension, then $c(E) = \{c(A) \mid A \in E\}$ is the correspodning *theory extension*.

An extension of an argumentation system yields a theory of the underlying logic under the conclusion function, a subset of formulas from $\mathcal{L}$. Such theories correspond to the theory extensions familiar from default logic – now obtained from argument extensions using the lifted conclusion function. Hence, a basic sanity check for structured argumentation asks whether the resulting theory extensions are always consistent theories. In [3], we find several rationality postulates for structured argumentation, including consistency of theory extensions. In this article, it is the only rationality postulate that we will be concerned with. This is justified by the fact that in systems like ASPIC$^+$, which satisfy conditions (2)-(9), it is the only non-trivial rationality postulate.

To formalise these remarks, for all $E \subseteq \mathcal{A}$ we lift notation and let $Sub(E) = \{A \in \mathcal{A} \mid \exists B \in E : A \in Sub(B)\}$ denote the sets of subarguments of a set of arguments E. Furthermore, for all $A \in \mathcal{A}$ let $\mathsf{Sc}(A) = \{B \in \mathcal{A} : A \in Sub(B) \text{ and } F_S(A) = F_S(B)\}$ be the set of *strict continuations* of argument A. We lift this notation to sets $E \subseteq \mathcal{A}$ as well, such that $\mathsf{Sc}(E) = \{A \in \mathcal{A} \mid \exists B \in E : A \in \mathsf{Sc}(B)\}$. Then the rationality postulates from [3] can be defined as follows.

Definition 5. Given an argumentation semantics ε and an argumentation system A, the *rationality postulates* for ε at A are the following:

  i. Subargument closure: $\forall E \in \varepsilon(\mathsf{A}) : Sub(E) \subseteq E$.

  ii. Strict closure: $\forall E \in \varepsilon(\mathsf{A}) : \mathsf{Sc}(E) \subseteq E$.

  iii. Direct consistency: $\forall E \in \varepsilon(\mathsf{A}) : \forall \phi \in c(E) : \overline{\phi} \cap c(E) = \emptyset$.

  iv. Indirect consistency: $\forall E \in \varepsilon(\mathsf{A}) : \forall \phi \in c(E) : c(E) \not\models_{\mathcal{L}} \neg\phi$ where $\models_{\mathcal{L}}$ is the logical consequence relation of the underlying logic for $\mathcal{L}$.

It is easy to see that first constraint follows from assumptions $(2) - (9)$ (in particular, by repeated applications of (4)). The second constraint then follows trivially for all systems that also satisfy (1), since strict continuations cannot make an argument more vulnarable to attack in this case. It follows that for systems like ASPIC$^+$, consistency is the only substantive rationality constraint. By contrast, when working with systems that violate (1), satisfying the closure properties can be non-trivial [4, 9]. This challenge is not discussed further in this work, where we focus on the question of consistency. As to the two consistency requirements, we focus only on direct consistency, since the indirect variant will be ensured whenever we have a sufficient set of strict rules available. In particular, indirect consistency follows from direct consistency and strict closure whenever the following is assumed (where $\perp$ is used to denote an arbitrary contradiction in the underlying logic):

$$\forall E \subseteq \mathcal{A} : ((c(E) \not\models_{\mathcal{L}} \perp \text{ and } c(E) \models_{\mathcal{L}} \phi) \Rightarrow \exists B \in \mathsf{Sc}(E) : c(B) = \phi) \qquad (10)$$

This is effectively a constraint on the deductive strength of the strict rules present in the system, telling us that A does indeed match the notion of logical consequence given by $\models_{\mathcal{L}}$, on all sets of arguments with mutually consistent conclusions. Unless such a condition is fulfilled, it seems inaccurate to say that $\models_{\mathcal{L}}$ is the "underlying" logic of A. If some logical consequences of arguments with mutually consistent conclusions cannot be argued for, it would be more appropriate to say that the underlying logic is a proper fragment of $\models_{\mathcal{L}}$. In this case, we also undercut the conceptual argument in favour of indirect consistency with respect to $\models_{\mathcal{L}}$. There is no reason why we would generally expect or require indirect consistency of logical consequences that it is not possible to argue for. Hence, when strict closure holds, indirect consistency is conceptually redundant. Since we only address systems satisfying conditions (2)-(9), our work in the following targets only the property of direct consistency, which we refer to from now on simply as *consistency*.

We remark that there are additional constraints sometimes grouped together with the original rationality postulates discussed above, most notably so-called crash-resistance and non-interference [5, 13, 9]. These are two closely related constraints that effectively restrict the use of *ex falso* as a strict rule when building arguments. The intuition behind such restrictions is to prevent the use of inconsistent fallible subarguments as means to successfully attack arbitrary unrelated arguments. Argumentation frameworks based on classical logic typically fail to exclude such vacuous attacks, since the unrestricted use of *ex falso* will often (depending on the preference ordering) allow us to defeat reasonable arguments using completely unrelated inconsistencies found among other default rules in the system.

While we do not investigate the formal details, it should be noted that crash-resistance and non-interference can be imposed as additional constraints, adding to (2)-(9) and (10) above. Notice that (10) only requires us to include enough strict rules to deduce the consequences of *consistent* sets of arguments. As long as closure and consistency holds, this means that strict rules corresponding to *ex falso* are not required. Hence, the constraints we rely on in this article are compatible with the assumption that non-interference and crash-resistance also hold, provided we are working with an argumentation system that satisfies (direct) consistency.

Therefore, the key question that concern us in this article is the following: how can we ensure that consistency is satisfied? The most obvious way to do it is to ensure that $B \in \mathsf{Def}(A)$ or $A \in \mathsf{Def}(B)$ whenever $c(A) \in \overline{c(B)}$. In this case, conflict-freeness ensures that there is no extension with both A and B, meaning that the corresponding theory extension cannot be inconsistent. However, this approach to the consistency problem makes the notion of strict reasoning rather vacuous – effectively, all reasoning steps become defeasible, regardless of whether we use a strict or a defeasbile rule.[2] Instead of this reductive solution, Prakken proposed a more subtle constraint for ensuring consistency [12]. We will present an abstract variant of his condition below.

3.2 Reasonableness and conflict contraposition

For all arguments $A \in \mathcal{A}$, we define $[A]_{\mathcal{R}_d} = \{B \in \mathcal{A} \mid \mathcal{R}^S_d(A) = \mathcal{R}^S_d(B)\}$. That is, $[A]_{\mathcal{R}_d}$ is an equivalence class of arguments containing exactly the same salient defeasible rules as A. Intuitively, these are the arguments that we regard as similar to A in an argumentation-theoretic sense. We will assume that when we reason contrapositively to defeat A, we will never need to use a similar argument to A in order to do so. That is, we will not need to use an argument that includes exactly the same set of defeasible rules as those that occur in A (we might well have to use all but one). Formally speaking, this strengthens the original definition of reasonableness, but if contrapositive reasoning is implemented in a reasonable way, the definitions will amount to equivalent restrictions. It should be noted that we need this adaptation in the proof of Theorems 2 and 3, but not elsewhere.

We will now state our version of Prakken's original reasonableness constraint. Our version is more general, as it does not involve preferences at all. However, we show in Proposition 1 that the original preference-dependent formulation implies reasonableness as defined below.

Definition 6 (Reasonableness – without preferences). For any argumentation system A, we say that A is reasonable if for all $A, B \in \mathcal{A}$ such that $c(A) \in \overline{c(B)}$,

---

[2]We mention that the weakest link system in [10] achieves consistency in a similar manner, by not including any strict rules. As we will show later, in Section 5, the approach to preferences found in [10] does not generalise well when strict rules are included and treated in the manner of ASPIC+.

$A \notin \mathsf{Def}(B)$ and $B \notin \mathsf{Def}(A)$ we have $\exists X \in M_S(A) \cup M_S(B)$:

$$\exists Y \in \mathcal{A} : X \in \mathsf{Def}(Y) \text{ and } M_S(Y) \subseteq (M_S(A) \cup M_S(B)) \setminus [X]_{\mathcal{R}_d}$$

That is, a system is reasonable if for any A, B with inconsistent conclusions, if A and B are unrelated by the relation of defeat, then there is a maximally fallible subargument, $X \in M_S(A) \cup M_S(B)$, that is defeated by an argument containing at most the fallible subarguments that already occur in A or B and that are not similar to X. Returning to the Dung example, we see that this fails, since no maximally fallible subargument (consisting of one of the r-rules) of any of the dilemmas could be defeated (since they were all strictly preferred to the arguments that attacked them). If reasonableness holds, it is not hard to prove that it does ensure consistency, as follows.

Theorem 1. *For any argumentation system* A, *if* A *is reasonable then it is consistent.*

Proof. Assume towards contradiction that A is inconsistent. Then there is some $E \in \varepsilon(\mathcal{A}, \mathsf{Def})$ with $\phi \in c(E)$ such that $c(E) \cap \overline{\phi} \neq \emptyset$. Say $\phi = c(B)$ and let $c(A) \in c(E) \cap \overline{\phi}$. Then $c(A) \in \overline{c(B)}$.

Since ε is conflict-free, it follows that $A \notin \mathsf{Def}(B)$ and $B \notin \mathsf{Def}(A)$. Hence, since A is reasonable, there is some $X \in M_S(A) \cup M_S(B)$ such that there is $Y \in \mathcal{A}$ with $X \in \mathsf{Def}(Y)$ and $M_S(Y) \subseteq (M_S(A) \cup M_S(B)) \setminus [X]_{\mathcal{R}_d}$. Assume wlog that $X \in M_S(A)$. Then by (4) it follows that $A \in \mathsf{Def}(Y)$. Hence, $Y \notin E$ by conflict-freeness of E. By completeness of ε, it follows that there is some $Z \in E$ such that $Y \in \mathsf{Def}(Z)$. By (4) there is some $V \in F_S(Y)$ such that $V \in \mathsf{Def}(Z)$. By (7) there are then two cases:

i. $V \in F_S(A)$. In this case, $A \in \mathsf{Def}(Z)$ with $A, Z \in E$.

ii. $V \in F_S(B)$. In this case, $B \in \mathsf{Def}(Z)$ with $B, Z \in E$.

Either way, we contradict conflict-freeness of ε. $\qquad\square$

In this article, we focus on the order-theoretic assumptions needed to ensure reasonableness for preference-dependent argumentation. Hence, the details of how contrapositive reasoning is done is of no particular interest to us. We will simply limit attention to argumentation systems where a sufficient number of contrapositive arguments are ensured to exists.

Definition 7 (Conflict contraposition). For any argumentation system A and all $A, B \in \mathcal{A}$, we say that A admits conflict contraposition if $c(A) \in \overline{c(B)}$ implies that $\forall X \in M_S(A) \cup M_S(B)$:

$$\exists Y \in \mathcal{A} : X \in \mathsf{Att}(Y) \text{ and } M_S(Y) \subseteq (M_S(A) \cup M_S(B)) \setminus [X]_{\mathcal{R}_d}$$

That is, conflict contraposition is the same requirement as reasonableness, except that X is under the scope of universal quantification while the relation of defeat is substituted by the (weaker) relation of attack. Clearly, for argumentation systems such that $\mathsf{Att}(A) = \mathsf{Def}(A)$ for all $A \in \mathcal{A}$, conflict contraposition implies reasonableness. Indeed, conflict contraposition is then a stronger requirement. It might be possible to use weaker requirements here, e.g., the notion of self-contradiction found in [8].[3] By assuming a strong and simple constraint that provides enough attacks we can focus on the order-theoretic aspect of the consistency problem: how to ensure that reasonableness is not violated when the notion of attack is pruned by a preference relation to yield a distinct notion of defeat. This is the subject we address in the next section.

4 Preferential systems and lifting principles

The intuition behind introducing preferences is simple enough: if A attacks B on some fallible subargument $B' \in F_S(B)$, the attack could fail if we strictly prefer B' over A. The paradigmatic example where this is reasonable is the situation where the attack is symmetric: A attacks B', but B' also attacks A. In this case, it makes sense that the choice comes down to our preferences. To characterise preference-dependent argumentation, we will use the following constraint.

Definition 8 (Preferential systems). An argumentation system A is preferential for the preorder $\preceq\, \subseteq \mathcal{A} \times \mathcal{A}$ if for all $A, B \in \mathcal{A}$:

$$B \in \mathsf{Att}(A) \text{ and } A \not\prec B \Rightarrow B \in \mathsf{Def}(A)$$

That is, in preferential argumentation systems we know that if A attacks B and B is not strictly preferred to A, then A also defeats B (notice that preference-independent attacks are not excluded; it is permitted for A to attack and defeat B even if B is preferred over A). Taken together with (4), it follows that if A attacks

---

[3]We remark that while [8] provides an abstract logic approach to the consistency problem, it does not account for preferences. Hence, while there are some similarities with our approach (most notably the level of generality achieved by abstracting away from how exactly arguments are built), we focus on different technical and conceptual challenges.

B' and B' is a fallible subargument of B, then A defeats B whenever B' is not strictly preferred to A. We also require the following natural property, stating that it is impossible to strengthen an argument by adding defeasible rules.

$$A \preceq B \text{ and } \mathcal{R}_d(A) \subseteq \mathcal{R}_d(C) \Rightarrow C \preceq B \tag{11}$$

With preferences in place, we can formulate a condition that ensures reasonableness, corresponding more closely to the original definition of reasonableness given by Prakken [12].

Proposition 1 (Reasonableness – with preferences). *Let A be an arbitrary argumentation system that admits conflict contraposition and is preferential for $\preceq$. Assume that for all $X, Y \in \mathcal{A}$ such that $c(Y) \in \overline{c(X)}$, $X \notin \mathsf{Def}(Y)$ and $X \prec Y$, there is $X_i \in M_S(X)$ and $Z \in \mathcal{A}$ such that:*

$$X_i \in \mathsf{Def}(Z) \text{ and } M_S(Z) \subseteq (M_S(X) \cup \{Y\}) \setminus [X_i]_{\mathcal{R}_d}$$

Then A is reasonable.

Proof. We consider arbitrary $A, B \in \mathcal{A}$ with $c(B) \in \overline{c(A)}$, $A \notin \mathsf{Def}(B)$ and $B \notin \mathsf{Def}(A)$. There are three cases:

  i. $A \prec B$. Then the claim follows immidiately from the assumption.

  ii. $B \prec A$. We have $B \notin \mathsf{Def}(A)$ and $c(A) \in \overline{c(B)}$. Hence, by assumption there is $B_i \in M_S(B)$ and $Z \in \mathcal{A}$ such that:

  $$B_i \in \mathsf{Def}(Z) \text{ and } M_S(Z) \subseteq (M_S(B) \cup \{A\}) \setminus [B_i]_{\mathcal{R}_d}$$

  This suffices to ensure reasonableness.

  iii. B and A are equivalent under $\preceq$. In this case, it is easy to see that (6) implies that $\mathsf{LastRule}(A), \mathsf{LastRule}(B) \in \mathbb{S}$. Let C be some element of $M_S(A) \cup M_S(B)$. By (9), $M_S(A) \cup M_S(B) \neq \emptyset$, so such C must exist. Assume wlog that $C \in M_S(A)$. By conflict contraposition, there is some $Y \in \mathcal{A}$ with $Y \in \overline{c(C)}$ and $M_S(Y) \subseteq (M_S(A) \cup M_S(B)) \setminus [C]_{\mathcal{R}_d}$. If $C \in \mathsf{Def}(Y)$, we are done, so assume $C \notin \mathsf{Def}(Y)$. If $Y \in \mathsf{Def}(C)$, then by (4) there must be some $D \in F_S(Y) \subseteq \{F_S(X) \mid X \in (M_S(A) \cup M_S(B)) \setminus [C]_{\mathcal{R}_d}\}$ (where the final inclusion follows from (7)). Moreover, $D \in \mathsf{Def}(C)$ by (4), so we are done. Hence, we may assume $C \notin \mathsf{Def}(Y)$ and $Y \notin \mathsf{Def}(C)$. But we know that $\mathsf{LastRule}(C) \in \mathbb{D}$ (since $C \in M_S(A)$), so by case i. and ii. above, combined with (4), reasonableness follows. $\square$

This proposition explains how reasonableness with preferences works, as a special case of the more general reasonableness property from Definition 6. Whenever X and Y are in conflict, it is safe to assume that when one of these arguments is strictly preferred over the other, we can use the strictly preferred argument to attack a fallible subargument of the other. To ensure consistency, choosing the strictly preferred argument in this way is not always necessary, as reflected by the formulation in Definition 6. Hence, Prakken's original definition of reasonableness includes an additional constraint motivated by (reasonable) intuitions about preferences, not strictly needed to ensure consistency. Proposition 1 shows how Prakken's formulation still suffices to ensure reasonableness in the more general sense.

Where do preferences over arguments come from? One possible answer, typically assumed in the theory of structured argumentation, is that our relative preference for an argument should match our relative preference (or priority) assigned to rules that occur in that argument.[4]

Recall that $\mathcal{R}_d^S(A)$ denotes the set of salient defeasible rules that occur in A. To formalise the idea that preferences for arguments should be possible to explain in terms of rule-orderings, we first define rule-based preferences, orderings over arguments that correspond to orderings over *sets* of rules.

Definition 9 (Rule-based preferences). An argumentation system A that is preferential for $\preceq$ has rule-based preferences if there is some relation $\trianglelefteq \subseteq 2^{\mathbb{D}} \times 2^{\mathbb{D}}$ such that for all $A, B \in \mathcal{A}$:

$$A \preceq B \Leftrightarrow \mathcal{R}_d^S(A) \trianglelefteq \mathcal{R}_d^S(B)$$

In this case we say that $\preceq$ is induced by $\trianglelefteq$.

We are now ready to define an abstract rationality constraint, which will ensure consistency for all argumentation systems with rule-based preferences and conflict contraposition. We remark that the constraint is not independently justified – as far as we can see, it admits no obviously desirable intuitive reading. However, it is both useful and desirable in view of how it provides a purely order-theoretic perspective of what consistency amounts to for the class of argumentation systems studied in this article. As such, it is used here to express rather surprising and clarifying results about the relationship between consistency and properties of rule-orderings.

Definition 10 (Rationality). A relation $\trianglelefteq \subseteq 2^{\mathbb{D}} \times 2^{\mathbb{D}}$ is *rational* at $\Delta \subseteq 2^{\mathbb{D}}$ if it satisfies the following constraint:

$$\Delta \neq \emptyset \Rightarrow \exists A \in \Delta : \bigcup_{B \in \Delta \setminus \{A\}} B \ntrianglelefteq A$$

---

[4]The rule-ordering, meanwhile, is typically taken as an exogeneous primitive: it is determined by the modeller, not the model.

If $\unlhd$ is rational at all $\Delta \subseteq 2^{\mathbb{D}}$, we say that $\unlhd$ is rational.

Proposition 2. *Let* A *be an argumentation system with conflict contraposition and rule-based preferences induced by* $\unlhd$. *Then* A *is consistent if for all* $A, B \in \mathcal{A}$ *such that* $c(A) \in \overline{c(B)}$, $A \notin \mathsf{Def}(B)$ *and* $A \prec B$, $\unlhd$ *is rational at* $\{\mathcal{R}_d^S(X) \mid X \in M_S(A) \cup \{B\}\}$.

Proof. We will show that A is reasonable using Proposition 1, so that the claim follows from Theorem 1. Consider arbitrary $A, B \in \mathcal{A}$ with $B \in \overline{c(A)}$, $A \notin \mathsf{Def}(B)$ and $A \prec B$. Let $\Delta = \{\mathcal{R}_d^S(X) \mid X \in M_S(A) \cup \{B\}\}$. That is, Δ includes exactly those sets of rules that are the salient defeasible rules of B or some maximally fallible subargument of A. By (9) we have $\Delta \neq \emptyset$. Hence, by rationality of $\unlhd$ there is some $X \in \Delta$ such that $\bigcup_{Y \in \Delta \setminus X} Y \ntrianglelefteq X$. Assume wlog that $\mathcal{R}_d^S(A_i) = X$ for some $A_i \in M_S(A)$. Since A admits conflict contraposition, there is some argument C with $A_i \in \mathsf{Att}(C)$ and $M_S(C) \subseteq (M_S(A) \cup M_S(B)) \setminus [A_i]_{\mathcal{R}_d}$. Hence, $\mathcal{R}_d^S(C) \subseteq \bigcup_{Z \in (M_S(A) \cup M_S(B)) \setminus [A_i]_{\mathcal{R}_d}} \mathcal{R}_d^S(Z) \subseteq \bigcup_{Y \in \Delta \setminus \mathcal{R}_d^S(A_i)} Y$, where the final inclusion relies on the fact that $\forall W \in M_S(A) \cup M_S(B) : \mathcal{R}_d^S(W) = \mathcal{R}_d^S(A_i) \Rightarrow W \in [A_i]_{\mathcal{R}_d}$. By (11) it follows that $\mathcal{R}_d^S(C) \ntrianglelefteq \mathcal{R}_d^S(A_i)$. Hence, it follows by Definition 8 that $A_i \in \mathsf{Def}(C)$ as desired. $\square$

The rationality constraint in Definition 10 is rather abstract. In the following, we will work with rule-based preferences of the elitist variety and show that rationality follows from some more familiar and intuitive order-theoretic constraints. This will also help shed more light on the intuitive content of the rationality constraint.

Definition 11 (Elitist lifting). A rule-based argumentation system A with preferences induced by $\unlhd$ is *elitist* if there is some $\leq\, \subseteq \mathbb{D} \times \mathbb{D}$ such that for all $R_1, R_2 \subseteq 2^{\mathbb{D}}$:

$$R_1 \unlhd R_2 \Leftrightarrow \begin{cases} R_2 = \emptyset \text{ if } R_1 = \emptyset \\ \exists r_1 \in R_1 : \forall r_2 \in R_2 : r_1 \leq r_2 \end{cases}$$

In this case, we say that $\unlhd$ is *lifted from* $\leq$.

The *strict* elitist lifting is obtained from replacing $r_1 \leq r_2$ by $r_1 < r_2$ in the definition above and removing the clause for $R_1 = \emptyset$. To sum up where we stand at this point, a rule-based argumentation system with an elitist preference ordering admits a characterisation in terms of three difference relations, for all $A, B \in \mathcal{A}$:

$$A \preceq B \Leftrightarrow \mathcal{R}_d^S(A) \unlhd \mathcal{R}_d^S(B) \Leftrightarrow \begin{cases} \mathcal{R}_d^S(B) = \emptyset \text{ if } \mathcal{R}_d^S(A) = \emptyset \\ \exists r_1 \in \mathcal{R}_d^S(A) : \forall r_2 \in \mathcal{R}_d^S(B) : r_1 \leq r_2 \end{cases}$$

Dung's example shows that elitist systems are not necesssarily rational. However, it is easy to show that strict elitist systems are rational for all preorders $\leq \subseteq \mathbb{D} \times \mathbb{D}$. This observation corresponds to the modified result on reasonableness in the corrected version of [11]. To see why it is true, consider arbitrary non-empty $\Delta \subseteq 2^{\mathbb{D}}$ and let r be a $\leq$-minimal element in $\bigcup \Delta$. Hence, there is no set in Δ that contains a rule strictly weaker than r. Let $R \in \Delta$ be some set containing r. Clearly, $\bigcup_{Y \in \Delta \setminus R} Y \ntrianglelefteq R$ under strict elitist lifting, since otherwise there would have to be a rule in $\bigcup_{Y \in \Delta \setminus R} Y$ that is strictly weaker than r.

As mentioned in Section 2, strict elitism is conceptually problematic. In the following, we will show that non-strict elitism does ensure rationality (at all Δ) under additional order-theoretic constraints. For all $A, B \in 2^{\mathbb{D}}$, let $A \perp B$ denote that $A \ntrianglelefteq B$ and $B \ntrianglelefteq A$, i.e., A and B are incomparable. For single rules, under elitist lifting, we have $r_1 \leq r_2 \Leftrightarrow \{r_1\} \trianglelefteq \{r_2\}$, so we need not distinguish between $X \perp r$ and $X \perp \{r\}$. The main result of this section is the following.

Theorem 2. *Let* A *by a rule-based argumentation system with conflict contraposition and preferences $\preceq$ induced from $\trianglelefteq$, lifted from $\leq$. Then* A *is consistent if one or more of the following conditions hold.*

    i. $\leq$ is a total relation over rules.

    ii. $\trianglelefteq$ is an antisymmetric relation over sets of rules.

    iii. $\trianglelefteq$ satisfies transitivity of incomparability, i.e., for all $A, B, C \in 2^{\mathbb{D}}$, if $A \perp B$ and $B \perp C$, then $A \perp C$.

Proof. By Proposition 2, it suffices to show that $\trianglelefteq$ is rational if any of the constraints (i)-(iii) hold. To prove these three claims, we first assume towards contradiction that there is some $\Delta \subseteq 2^{\mathbb{D}}$ with $\Delta \neq \emptyset$ such that $\bigcup_{Y \in \Delta \setminus A} Y \triangleleft A$ for all $A \in \Delta$. Let $b \in \bigcup_{Y \in \Delta \setminus A} Y$ be minimal such that $\forall a \in A : b \leq a$. Since $\trianglelefteq$ is lifted from $\leq$, such b must exist. Then we have the following:

- There is some $B \in \Delta \setminus A$ such that $b \in B$. By assumption, we also have $\bigcup_{Y \in \Delta \setminus B} Y \triangleleft B$, with some $c \in \bigcup_{Y \in \Delta \setminus B} Y$ as a minimal witness with $\forall x \in B : c \leq x$. Specifically, we have $c \leq b$. By transitivity, we then also have $c \leq a$ for all $a \in A$. Since b was chosen minimally, it follows that $b \equiv c$. Notice that $b = c$ is possible (this prevents us from proving stronger results).

- Since $B \ntrianglelefteq \bigcup_{Y \in \Delta \setminus B} Y$, there must be $d \in \bigcup_{Y \in \Delta \setminus B} Y$ such that $b \equiv c \nleq d$. By minimality of $b \in B \subseteq \bigcup_{Y \in \Delta \setminus A} Y$, it follows that $b \perp d$. This establishes point i of the claim, since it contradicts the assumption that $\leq$ is total.

- By transitivity of $\leq$ and $b \equiv c$ it also follows that $c \bot d$. Since $c \in \bigcup_{Y \in \Delta \setminus B} Y$, there is $C \in \Delta \setminus \{B\}$ with $c \in C$. Moreover, $C \neq B$. By assumption, we have $\bigcup_{Y \in \Delta \setminus C} Y \lhd C$, with some $e \in \bigcup_{Y \in \Delta \setminus C} Y$ that is a minimal witness with $\forall x \in C : e \leq x$. It follows that $e \leq c$, so that by minimality of c we get $e \equiv c$. By now, we have $e \equiv c \equiv b$ (but they could all be the same rule).

- Since $C \ntrianglelefteq \bigcup_{Y \in \Delta \setminus C} Y$, we obtain $f \in \bigcup_{Y \in \Delta \setminus C} Y$ such that $e \bot f$. We then get two cases:

  - $e \neq c$. Then $e \equiv c$ violates antisymmetry (since then $\bigcup_{Y \in \Delta \setminus B} Y \trianglelefteq \bigcup_{Y \in \Delta \setminus C} Y, \bigcup_{Y \in \Delta \setminus C} Y \trianglelefteq \bigcup_{Y \in \Delta \setminus B} Y$ with $\bigcup_{Y \in \Delta \setminus B} Y \neq \bigcup_{Y \in \Delta \setminus C} Y$), so point ii of the claim follows. Moreover, we have $e \bot f \bot c$ with $e \equiv c$, so incomparability is not transitive and point iii of the claim follows as well.

  - $e = c$. Since $C \neq B$, this means that either $C \neq \{c\}$ or $B \neq \{c\}$. Assume wlog that $C \neq \{c\}$. Then $C \equiv \{c\}$ violates antisymmetry, so point ii of the claim follows. Moreover, we have $C \bot f \bot c$ with $C \equiv \{c\}$, so incomparability is not transitive and point iii of the claim follows as well.

$\square$

It is instructive to notice how Dung's example, considered in Section 2, violated all three conditions for rationality identified above. First, $\leq$ was clearly not a total relation, since the rules were partitioned into incomparable equivalence classes. Second, $\trianglelefteq$ is not antisymmetric, since $A_1 \equiv A_2$ and $A_3 \equiv A_4$, Finally, transitivity did not hold for incomparability under $\trianglelefteq$, since $A_1 \bot A_3 \bot A_2$, even though A_1 and A_2 were equivalent.

More generally, notice that point ii of Theorem 2 implies that consistency holds for strict elitist lifting. Indeed, it is not hard to see that if strict elitist lifting is used, then $\trianglelefteq$ will be antisymmetric over sets of rules; only $\emptyset$ is equivalent to anything (namely itself) under strict elitist lifting (in [11], $A \equiv B$ when $A = B$ is added by a special clause, to make the set-comparison reflexive, but this does not violate antisymmetry).

On the other hand, notice that antisymmetry of $\leq$ does not imply antisymmetry of $\trianglelefteq$. For example, A and B can share a weakest rule r such that $A \trianglelefteq B, B \trianglelefteq A$ even though $A \neq B$. Similarly, transitivity of incomparability with respect to $\trianglelefteq$ does not imply transitivity of incomparability with respect to $\leq$. To see this, consider a collection of six rules $r_1, \ldots, r_6$ such that $r_1 < r_2 < r_3$ and $r_4 < r_5 < r_6$, while all other pairs of distinct rules are incomparable. Then incomparability under $\leq$ is transitive, as the reader can verify. However, consider the sets $A_1 = \{r_1, r_2\}$, $A_2 = \{r_1, r_3\}$ and $A_3 = \{r_4, r_5\}$, $A_4 = \{r_4, r_6\}$. Under elitist lifiting, these sets will be ordered exactly as the problematic rules were ordered in the Dung example:

$$
\begin{array}{ccc}
A_1 & \equiv & A_2 \\
\perp & & \perp \\
A_3 & \equiv & A_4
\end{array}
$$

We see that $A_1 \perp A_3 \perp A_2$, even though $A_1 \equiv A_2$, witnessing to the fact that incomparability at the set level is not transitive. Moreover, it is easy to verify that $\trianglelefteq$ induced by $\leq$ under elitist lifting is not rational at $\Delta = \{A_1, A_2, A_3, A_4\}$. Hence, stipulating antisymmetry and/or transitivity of incomparability with respect to $\leq$ does *not* guarantee rationality. We can easily instantiate the rules above in such a way that an inconsistency arises, showing that these two conditions are not in any case strong enough to give us what we want when restricted to $\leq$. However, as we will see in Section 6, an additional assumption on $\leq$ will allow us to replace $\trianglelefteq$ by $\leq$ in all points of Theorem 2. Before getting to this result, we will address a conceptual problem, threatening to undermine the very idea of elitist lifting.

When attention is restricted to argumentation systems with last link salience, this changes. Then we may replace $\trianglelefteq$ by $\leq$ in all points of Theorem 2, as we will demonstrate in the following.

First we need to make an additional assumption regarding the nature of the contrapositive arguments we have available. There are several ways of getting what we need, including strengthening the definition of conflict contraposition to ensure that we never have to use r to attack A when $M_S(A) = \{A\}$ and $\mathsf{LastRule}(A) = r$. For simplicity, we choose a different approach here, whereby we assume that no two maximally fallible subarguments of the same argument conclude with the same defeasible rule. This is not a very limiting assumption, depending on the exact format of the strict rules we have available. Specifically, assuming a Hilbert-style representation of arguments as sequences of rules, the assumption is clearly inconsequential (since there is no reason for any rule to appear twice in such a sequence), c.f., [10]. If arguments have a tree-structure, by contrast, the same rule can appear in different positions in the tree. However, if two maximally fallible subarguments conclude with the same rule, it is semantically reasonable (in systems like ASPIC$^+$, at least) to remove either one of them from the set of maximally fallible subarguments.[5] Formally, we assume that the following property holds, for

---

[5]Roughly, the reason is that if A includes two maximally fallible subargument, B_1 and B_2, such that $\mathsf{LastRule}(B_1) = \mathsf{LastRule}(B_2) = r$, then the system also includes an argument A_1 (resp. A_2) that is identical with A except for the fact that B_2 has been replaced with B_1 (resp. B_1 has been replaced with B_2). But then the existence of suitable contrapositive arguments for both A_1 and

all $A \in \mathcal{A}$:

$$\forall A \in \mathcal{A} : \forall B_1, B_2 \in M_S(A) : B_1 \neq B_2 \Rightarrow \mathsf{LastRule}(B_1) \neq \mathsf{LastRule}(B_2) \qquad (12)$$

It is important to notice that the condition should not be strengthened in the obvious way, by requiring that no two distinct maximally fallible subarguments share a rule. Consider, for instance, a system where any argument needs to begin with the same initial rule r, the only rule that has an empty set of premises. Here we cannot assume to be working with disjoint sets of arguments; distinctness of final rules is all we have to rely on. However, it suffices to strengthen Theorem 2 for last link lifting.

Let $LL(A) = \{\mathsf{LastRule}(A_i) \mid A_i \in M_S\}$ and recall that last link salience obtains just in case $\mathcal{R}_d^S(A) = LL(A)$ for all $A \in \mathcal{A}$. Then we have the following.

Theorem 3. *Assume that* A *is a rule-based argumentation system satisfying (12), with last link salience, conflict contraposition and preferences* $\preceq$, *induced from* $\trianglelefteq$ *and lifted from* $\leq\ \subseteq \mathbb{D} \times \mathbb{D}$ *under elitist lifting. Then* A *is consistent if one or more of the following conditions hold.*

  i. $\leq$ *is antisymmetric.*

  ii. $\leq$ *satisfies transitivity of incomparability.*

Proof. The proof uses a similar construction as the proof of Theorem 2, but takes the structure of arguments into account. By Proposition 2, it suffices to show that $\trianglelefteq$ is rational for all $\Delta =$

$$\{\mathcal{R}_d^S(A_1), \ldots, \mathcal{R}_d^S(A_n), \mathcal{R}_d^S(B)\} = \{\{\mathsf{LastRule}(A_1)\}, \ldots, \{\mathsf{LastRule}(A_n)\}, LL(B)\}$$

where $\{A_i \mid 1 \leq i \leq n\} = M_S(A)$ for some $A, B \in \mathcal{A}$ such that $c(A) \in \overline{c(B)}$. Assume towards contradiction that rationality fails for some such Δ. As before, we let $\bigcup_{Y \in \Delta \setminus X} Y = \bigcup_{Y \in \Delta \setminus \{X\}} Y$ for all X. Hence, we have $\bigcup_{Y \in \Delta \setminus D} Y \lhd D$ for all $D \in \Delta$. Specifically, we have $\bigcup_{Y \in \Delta \setminus \mathcal{R}_d^S(B)} Y \lhd \mathcal{R}_d^S(B)$. We let $a \in \bigcup_{Y \in \Delta \setminus \mathcal{R}_d^S(B)} Y$ be minimal such that $\forall x \in \mathcal{R}_d^S(B) : a \leq x$. Then there is some $\mathcal{R}_d^S(A_i) \in \Delta \setminus \mathcal{R}_d^S(B)$ such that $a \in \mathcal{R}_d^S(A_i)$. Since A satisfies last link salience and $M_S(A_i) = \{A_i\}$, it follows that $a = \mathsf{LastRule}(A_i)$. By assumption, we also have $\bigcup_{Y \in \Delta \setminus \mathcal{R}_d^S(A_i)} Y \lhd \mathcal{R}_d^S(A_i)$, with some $c \in \bigcup_{Y \in \Delta \setminus \mathcal{R}_d^S(A_i)} Y$ as a minimal witness with $\forall x \in \mathcal{R}_d^S(A_i) : c \leq x$. Specifically, we have $c \leq a$. By transitivity, we then also have $c \leq x$ for all $x \in \mathcal{R}_d^S(B)$. Since a was chosen minimally, it follows that $a \equiv c$. There are two cases to consider.

A_2 ensures that we can drop either B_1 or B_2 from $M_S(A)$, without violating conflict contraposition at A.

i. $c \in \mathcal{R}_d^S(A_j)$ for some $A_j \in M_S(A)$. Then $b \neq c$ by (12), so $b \equiv c$ contradicts antisymmetry of $\leq$, proving point i. Moreover, $\mathcal{R}_d^S(A_j) \ntrianglelefteq \bigcup_{Y \in \Delta \setminus \mathcal{R}_d^S(A_j)} Y$ so there must be some $d \in \bigcup_{Y \in \Delta \setminus \mathcal{R}_d^S(A_j)} Y$ such that $c \nleq d$. Since c was chosen minimally and $c \equiv a$, it follows that $d \nleq c \equiv a$. Hence, $d \perp c$ and $d \perp a$, even though $c \equiv a$ and $c \neq a$. This contradicts transitivity of incomparability under $\leq$, establishing point ii of the claim.

ii. $c \in \mathcal{R}_d^S(B)$. Since $\mathcal{R}_d^S(B) \ntrianglelefteq \bigcup_{Y \in \Delta \setminus \mathcal{R}_d^S(B)} Y$ there is $d \in \bigcup_{Y \in \Delta \setminus \mathcal{R}_d^S(B)} Y$ such that $d \ngtr c$ (since otherwise c would witness to $\mathcal{R}_d^S(B) \trianglelefteq \bigcup_{Y \in \Delta \setminus \mathcal{R}_d^S(B)} Y$). However, c was chosen minimally from $\bigcup_{Y \in \Delta \setminus \mathcal{R}_d^S(B)} Y$ and $a = c$, so $d \ngtr c$ implies that $d \perp c$. Since we have $a = c \in \mathcal{R}_d^S(A_i) \cap \mathcal{R}_d^S(B)$ and $\forall x \in \mathcal{R}_d^S(A_i) \cup \mathcal{R}_d^S(B)$: $c = b \leq x$, it follows that $d \notin \mathcal{R}_d^S(A_i) \cup \mathcal{R}_d^S(B)$. Hence, there must be some $j \neq i$ with $d \in \mathcal{R}_d^S(A_j)$. By assumption, we have $\bigcup_{Y \in \Delta \setminus \mathcal{R}_d^S(A_j)} Y \triangleleft \mathcal{R}_d^S(A_j)$, so there is a minimal $e \in \bigcup_{Y \in \Delta \setminus \mathcal{R}_d^S(A_j)} Y$ with $\forall x \in \mathcal{R}_d^S(A_j) : e \leq x$. Since a was chosen minimally, we know that $e \nleq a = c$. Hence, $e \perp a = c$, since otherwise $c \leq d$ by transitivity of $\leq$, contradicting $c \perp d$. From this it follows that $e \notin \mathcal{R}_d^S(A_i) \cup \mathcal{R}_d^S(B)$, since $c = a$ is comparable to all elements of this set. This, in turn, implies that $e = \mathsf{LastRule}(A_k)$ for $i \neq k \neq j$. However, we also have $\bigcup_{Y \in \Delta \setminus \mathcal{R}_d^S(A_k)} Y \triangleleft \mathcal{R}_d^S(A_k)$, witnessed by some minimal $f \in \bigcup_{Y \in \Delta \setminus \mathcal{R}_d^S(A_k)} Y$ such that $f \leq x$ for all $x \in \mathcal{R}_d^S(A_k)$. Moreover, since $e \perp c$ and $c \leq x$ for all $x \in \mathcal{R}_d^S(A_i) \cup \mathcal{R}_d^S(B)$, it follows that $f \notin \mathcal{R}_d^S(A_i) \cup \mathcal{R}_d^S(B)$. Hence, we have $f = \mathsf{LastRule}(A_l)$ for some $l \neq k$. It follows by (12) that $f \neq e$. Since f was chosen minimally, we have $e \equiv f$, contradicting antisymmetry of $\leq$, thereby establishing point i. We also have $f \perp c \perp e$, contradicting transitivity of incomparability under $\leq$, thereby establishing point ii of the claim.

□

5 A conceptual problem and possible refinements

Assume you are faced with the following three rules pertaining to the evolution and creationism debate: r_1: $\Rightarrow c_1$, where c_1 is "evolution has not been proven"; r_2: $c_1 \Rightarrow c_2$, where c_2 is "we need more research on evolution"; and r_3: $c_1 \Rightarrow \neg c_2 \wedge c_3$, where c_3 is "the hypothesis of creationism should be entertained". We assume that the ordering over the rules is $r_1 < r_3 < r_2$.[6]

---

[6]Intuitively, saying that evolution has not been proven is quite a stretch, but not logically or even scientifically incoherent (depending on what you mean by "proven", obviously). However, if the scientist conceeds (as good scientists tend to do) that evolution has not been proven – in some

Let A be the argument obtained by taking r_1 followed by r_2, in favour of research, while B is the argument obtained by taking r_1 followed by r_3, against research and in favour of considering the hypothesis of creationism. Intuitively, A is a better argument than B, at least given our rule-ordering which makes clear that $r_2 > r_3$. Both arguments share r_1 as the least preferred rule, but r_2 is preferred over r_3. Still, the elitist lifting principle forces us to conclude that these two arguments are equally good, which seems intuitively unreasonable.

A possible solution, proposed in recent work [15, 10], is to use a *disjoint* version of elitist lifting. Specifically, we say that $\unlhd$ is a disjoint elitist lifting of $\leq$ whenever we have, for all $A, B \in 2^{\mathbb{D}}$:

$$A \unlhd B \Leftrightarrow \begin{cases} B = \emptyset \text{ if } A = \emptyset \\ \exists r_1 \in A \setminus B : \forall r_2 \in B \setminus A : r_1 \leq r_2 \text{ otherwise} \end{cases} \tag{13}$$

The strict variant of disjoint lifting is then defined in the obvious way, with $\leq$ replaced by $<$ in the equation above.

With disjoint lifting, the problem with the evolution debate above disappears. A is better than B since r_1, the weakest link of both arguments, is removed from consideration. There are two problems with this proposed solution, the first of which is conceptual. If we modify the evolution debate by including an additional rule $r_4 :\Rightarrow c_4$, where c_4 is "evolution could be incorrect", we can reasonably include also the rule r_5: $c_4 \Rightarrow \neg c_2 \wedge c_3$. The obvious extension of the rule-ordering would be $r_1 \equiv r_4 < r_3 \equiv r_5 < r_2$. The creationist could then form the argument C, built from r_4 followed by r_5. In this case, it would follow that A is equally good as C even under the disjoint lifting. Indeed, A and C share no rules, so disjoint lifting is the same as traditional elitist lifting in this case.

To solve this problem, we could try to expand upon the idea behind disjoint lifting, to remove not only shared rules between A and B but also rules belonging to shared equivalence classes of rules. Formally, we let $Eq(A) = \{r \in \mathbb{D} \mid \exists q \in A : r \equiv q\}$ and define strongly disjoint lifting as follows, for all $A, B \in 2^{\mathbb{D}}$:

$$A \unlhd B \Leftrightarrow \begin{cases} B = \emptyset \text{ if } A = \emptyset \\ \exists r_1 \in A \setminus Eq(B) : \forall r_2 \in B \setminus Eq(A) : r_1 \leq r_2 \end{cases} \tag{14}$$

Conceptually, strongly disjoint lifting seems like it might be exactly what we are after. Unfortunately, there is a second problem with the idea of removing shared rules from consideration: it yields inconsistency in many situations. For strongly

---

sense or other – they give the creationist an entry point to further their argument on a more solid footing, which would not have been afforded to creationism otherwise. This suggests the pragmatic correctness of the rule-ordering given here.

disjoint lifting, this is quite easy to see. For instance, the reader can check that strongly disjoint lifting results in inconsistency when applied to the Dung example in Section 2. In fact, contradictory theory extensions arise for even more basic examples, suggesting that strongly disjoint lifting is not a good solution to the conceptual problem discussed above.

Example 1 (Three rights make it wrong). We assume given an argumentation system with $\mathbb{D} = \{r_1, r_2, r_3\}$ and $\mathbb{S} = \{s_1, s_2, s_3\}$ where

$$\underbrace{\Rightarrow q_1}_{r_1}, \quad \underbrace{\Rightarrow q_2}_{r_2}, \quad \underbrace{\Rightarrow q_3}_{r_3}, \quad \underbrace{q_2, q_3 \to \neg q_1}_{s_1} \quad \underbrace{q_1, q_3 \to \neg q_2}_{s_2} \quad \underbrace{q_1, q_2 \to \neg q_3}_{s_3}$$

The argumentation system is preferential with an ordering obtained using strongly disjoint lifting of the rule-ordering $\leq \subseteq \mathbb{D} \times \mathbb{D}$ given by $r_1 \equiv r_3 < r_2$. Consider the following pairs of argument dilemmas:

$$\bar{A}_1 : \frac{\overset{\sim\sim r_2}{q_2}, \overset{\sim\sim r_3}{q_3}}{\neg q_1} s_1, \quad A_1 : \overset{\sim\sim r_1}{q_1}$$

$$\bar{A}_2 : \frac{\overset{\sim\sim r_1}{q_1}, \overset{\sim\sim r_3}{q_3}}{\neg q_2} s_2, \quad A_2 : \overset{\sim\sim r_2}{q_2}$$

$$\bar{A}_3 : \frac{\overset{\sim\sim r_1}{q_1}, \overset{\sim\sim r_2}{q_2}}{\neg q_3} s_3, \quad A_3 : \overset{\sim\sim r_3}{q_3}$$

Since $\preceq$ is obtained by strongly disjoint lifting from $\leq$, we have $\bar{A}_i < A_i$ for all $i \in \{1, 2, 3\}$. To see this, consider first the case when $i = 1$ and the case when $i = 3$. Either way, we have $Eq(\bar{A}_i) = \{r_1, r_2, r_3\}$. Hence, $A_i \setminus Eq(\bar{A}_i) = \emptyset$, so the strongly disjoint lifting definition (14) trivially ensures $\bar{A}_i \trianglelefteq A_i$ for these two cases. For $i = 2$, recall that $r_2 > r_3 \equiv r_1$, so taking either r_1 or r_3 as the witness shows that $\bar{A}_2 \trianglelefteq A_2$. Hence, any complete argumentation semantics will return all of $\{q_1, q_2, q_3\}$ as the extension, yielding an inconsistent theory (since all of $\neg q_i$ will then also have to be included, due to the strict rules).

As it turns out, the problem of consistency arises for both types of disjoint lifting, also for rule-orderings that are very well-behaved. Indeed, inconsistency arises already for linear orders, as shown by the following example.

Example 2. We let $\mathbb{D} = \{r_1, r_2, \ldots, r_6, q_1, q_2, q_3\}$ contain 9 defeasible rules. Each r_i is of the form $\Rightarrow p_i$ where $p_i \in \mathcal{L}$ is a propositional atom. That is, each r_i is a defeasible rule allowing you to conclude p_i from the empty premise. The three q-rules are depicted below:

$$\underbrace{p_1, p_3, p_6 \Rightarrow a_1}_{q_1}, \ \underbrace{p_1, p_2, p_5 \Rightarrow a_2}_{q_2}, \ \underbrace{p_2, p_4, p_5 \Rightarrow a_3}_{q_3}$$

So q_i allows you to conclude a_i from three p-premises, as detailed above. Furthermore, assume we have the following strict rules:

$$\underbrace{a_2, a_3 \to \neg a_1}_{s_1}, \underbrace{a_1, a_3 \to \neg a_2}_{s_2}, \underbrace{a_1, a_2 \to \neg a_3}_{s_3}$$

That is, an extension including $\{a_1, a_2, a_3\}$ will be inconsistent according to the underlying logic. The ordering $\leq \subseteq \mathbb{D} \times \mathbb{D}$ is given by $r_1 < r_2 < r_3 < r_4 < r_5 < r_6 < q_1 < q_2 < q_3$. That is, all q-rules are better than all r-rules and rules with a higher index are better than rules with a lower index. Consider the following pairs of argument dilemmas (names of rules are left implicit).

$$\bar{A}_3 : \frac{\dfrac{p_1, p_3, p_6}{a_1} \quad \dfrac{p_1, p_2, p_5}{a_2}}{\neg a_3}, \qquad A_3 : \frac{p_2, p_4, p_5}{a_3}$$

$$\bar{A}_2 : \frac{\dfrac{p_1, p_3, p_6}{a_1} \quad \dfrac{p_2, p_4, p_5}{a_3}}{\neg a_2}, \qquad A_2 : \frac{p_1, p_2, p_5}{a_2}$$

$$\bar{A}_1 : \frac{\dfrac{p_1, p_2, p_5}{a_2} \quad \dfrac{p_2, p_4, p_5}{a_3}}{\neg a_1}, \qquad A_1 : \frac{p_1, p_3, p_6}{a_1}$$

The argument pairs above show that disjoint elitist lifting does not satisfy the order-theoretic rationality constraint. Specifically, it is easy to check that we have $\bar{A}_i \prec A_i$ for all $i \in \{1, 2, 3\}$ for disjoint elitist lifting. Essentially, the reason is that the two subarguments of each $\bar{A}_i$-argument overlap with the weakest rules of A_i, so that removing shared rules forces the comparison to take place relative only to the strongest rules in A_i, which are all strictly preferred over a remaining rule of intermediate strength in $\bar{A}_i$. As a result, all complete extensions will include $\{a_1, a_2, a_3\}$, since all corresponding A_i-arguments are undefeated. This yields an inconsistent theory, when the strict rules are applied.

The example above shows that disjoint lifting can lead to violations of consistency for orderings we would normally consider very well-behaved, also for disjoint strict lifting. A minor modification to the example also answers in the negative a conjecture by Young and others, to the effect that disjoint lifting of any so-called *structure preference ordering* results in a rational ordering of arguments [14]. However, Young and others offer a direct proof that consistency is satisfied under disjoint lifting of such orderings (assuming enough strict rules are available to capture classical logic) [15]. This shows that structure preference orderings manage to avoid counterexamples even if they do not satisfy the purely order-theoretic rationality constraint.

To illustrate how this is possible, we first note that structure preference orderings forbid us from saying that q_1 and q_2 are better than any of the r-rules in the example above. The reason is that these two q-rules can only be applied after the weakest p-rule has already been applied. The reordering removes the inconsistency, but not the order-theoretic irrationality (or unreasonableness, in the terminology of [11]). To disprove the conjecture made by Young and others, we could either reorder or simply remove all q-rules. The sets of defeasible rules associated with the problematic arguments above would still exist and would still violate order-theoretic rationality. However, they would no longer result in any inconsistency.

In the next section, we will consider the phenomena at work here in more depth, by giving a more abstract (and much simpler) definition that captures the essential consistency-preserving property of the structure preference orderings defined by Young and others.

6   On structural orderings

In this section, we formulate an additional condition on arbitrary $\leq\ \subseteq \mathbb{D} \times \mathbb{D}$, capturing a key property of structure preference ordering in an abstract way. We also prove that every such ordering satisfies consistency under disjoint strict lifting, generalising previous results of Young and others. Specifically, we do not need to make any additional order-theoretic assumptions to prove the result.

Following up on this, we prove a result on traditional (non-strict) elitist lifting, strengthening Theorem 2 for the case when $\leq$ is a structural ordering. This is not only a technical result, but also conceptually significant. The reason is that the conceptual problems discussed in the previous section, motivating the shift to disjoint lifting, do not seem to arise when $\leq$ corresponds to a structure preference ordering.

Hence, it can be argued that disjoint lifting is a conceptual dead end: the solution to the problems discussed in the previous section is to reorder $\leq$ to take into account the structure of arguments. The solution is not to remove shared rules before comparing arguments. Indeed, Example 2 shows that disjoint lifting by itself does not ensure consistency. We also need to ensure that $\leq$ is structural. However, once we ensure this, the examples used to motivate disjoint lifting become inadmissible for an independent reason: the weakest link can no longer come at the beginning of two arguments that continue along a preferred and a strictly less preferred branch. Assuming all examples motivating disjoint lifting have this form, we might as well go back to traditional non-disjoint elitist lifting.

Formally, we start by making the following assumption, which is already satisfied by many argumentation systems, including those in the ASPIC^+ tradition.

$$\forall A \in \mathcal{A} : \forall B \in M_S(A) : \mathsf{LastRule}(B) \in \mathcal{R}_d^S(B) \subseteq \mathcal{R}_d^S(A) \tag{15}$$

That is, we assume that all maximally fallible rules of A conclude with a defeasible rule that is salient in both A and B. We are ready for the key definition of this section, whereby we will say that $\leq \subseteq \mathbb{D} \times \mathbb{D}$ is *structural* for $\mathcal{A}$ if

$$\forall A \in \mathcal{A} : \mathsf{LastRule}(A) \in \mathcal{R}_d^S(A) \Rightarrow \forall r \in \mathcal{R}_d^S(A) : r \not< \mathsf{LastRule}(A) \tag{16}$$

Effectively, structural rule-orderings take the structure of arguments into account, by ensuring that whenever a defeasible rule r occurs in an argument, all premises of r have been established using defasible rules that are at least as good as r. Intuitively, the fact that q can be used to establish a premise of r is a structural reason to regard q as a rule that is at least as good as r. This makes intuitive sense if we argue using a greedy heuristic, where we always use a maximally preferred rule to extend our arguments. For such a reasoner, a structural rule-ordering only encodes at the rule-level what already transpires at the argument level; rules are ordered (also) by the stage at which they may be applied (formally, this leads to a structural reordering of the original rule-order used by a greedy arguer, c.f., [15]).

Strictly speaking, the structural preference orderings defined by Young and others do not automatically ensure that (16) holds for every argument in $\mathcal{A}$. However, any argumentation system with a structural preference ordering can be characterised by a smaller collection of normal-form arguments that do satisfy the requirement. We will not formalise the construction from [15] to show this, but we will give a more general characterisation result below (Proposition 4) that clearly applies also to structural preference orderings. First, we show that (16) is all we need to ensure consistency under disjoint strict lifting.

Theorem 4. *Assume* A *is an argumentation system with conflict contraposition and an argument preference ordering that is induced by a set ordering that has been lifted under disjoint weakest link from a structural rule-ordering. Then* A *is consistent.*

Proof. We show that A is reasonable. Consider arbitrary A, B such that $c(A) \in \overline{c(B)}$, $A \notin \mathsf{Def}(B)$ and $B \notin \mathsf{Def}(A)$. We have to show $\exists X \in M_S(A) \cup M_S(B)$:

$$\exists Y \in \mathcal{A} : X \in \mathsf{Def}(Y) \text{ and } M_S(Y) \subseteq (M_S(A) \cup M_S(B)) \setminus [X]_{\mathcal{R}_d}$$

First, let $A_i \in M_S(A)$ such that (i) $\forall C \in M_S(A) : \mathsf{LastRule}(C) \not< \mathsf{LastRule}(A_i)$. That is, A_i is a maximally fallible subargument of A that concludes with a rule that has minimal priority of all rules that occur as the final rule in some maximally fallible subargument of A. Now, by (i) and the fact that $\leq$ is structural for A, it follows that (ii) $\forall r \in \mathcal{R}_d(A) : r \not< \mathsf{LastRule}(A_i)$. By conflict contraposition, we have an argument D that attacks A_i such that $\mathcal{R}_d^S(D) \subseteq \bigcup_{Z \in (M_S(A) \cup M_S(B)) \setminus [A_i]_{\mathcal{R}_d}} \mathcal{R}_d^S(Z)$. By (12), we know that $\mathsf{LastRule}(A_i) \notin \mathcal{R}_d^S(D)$. Hence, $\mathsf{LastRule}(A_i) \in \mathcal{R}_d^S(A_i) \setminus \mathcal{R}_d^S(D)$. Moreover, by (ii) we have $\forall r \in \mathcal{R}_d^S(D) \setminus \mathcal{R}_d^S(A_i) : r \not< \mathsf{LastRule}(A_i)$. Since $\trianglelefteq$ is lifted under disjoint strict lifting, it follows that $D \ntrianglelefteq A_i$. Hence, D defeats A_i as desired. $\qquad \square$

Notice that the proof does not hold when the lifting is not strict. In this case, it is possible for $A \preceq B$ with a witness $r \in \mathcal{R}_d(A)$ such that there is $q \in \mathcal{R}_d(B)$ with $r \equiv q$. Moreover, (12) does not rule out the possibility that there are $B_1, B_2 \in M_S(A)$ such that $\mathsf{LastRule}(B_1) \equiv \mathsf{LastRule}(B_2)$ with $\mathsf{LastRule}(B_1) \neq \mathsf{LastRule}(B_2)$ being two distinct minimally preferred rules occuring as the final defeasible rule applied in a maximally fallible subargument of A. Specifically, the Dung example is a counter-example to consistency, for both disjoint and ordinary non-strict lifting, since the rule-ordering from this example is structural.

However, when the rule-ordering is structural, the elitist principle is *almost* representable by last link reasoning, for any definition of salience ($\mathcal{R}_d^S$) that satisfies $\mathsf{LL}(A) \subseteq \mathcal{R}_d^S(A)$ for all $A \in \mathcal{A}$. Intuitively, this is because the weakest rules in an argument are always applied last when the ordering is structural. To make the last link representation tight, we have to strengthen our requirement, giving rise to what we call *strongly* structural orderings, defined as follows:

$$\forall A \in \mathcal{A} : \mathsf{LastRule}(A) \in \mathcal{R}_d^S(A) \Rightarrow \forall r \in \mathcal{R}_d^S(A) : r \leq \mathsf{LastRule}(A) \qquad (17)$$

If an ordering is strongly structural, last link reasoning is always equivalent to weakest link reasoning, meaning that Theorem 3 applies to both. To formalise this, say that A is last link representable if the following holds, for all $A, B \in \mathcal{A}$:

$$A \not\prec B \quad \Leftrightarrow \quad A = \emptyset \text{ or } \forall x \in \mathsf{LL}(A) : \exists y \in \mathsf{LL}(B) : x \not\leq y \qquad (18)$$

Intuitively, A is last link representable if its defeat relation can be equivalently characterised using last link lifting of $\leq$. Obviously, if A satisfies last link salience, so that $\mathcal{R}_d^S(A) = \mathsf{LL}(A)$ for all $A \in \mathcal{A}$, it is last link representable. Conversely, if A is last link representable, then replacing $\mathcal{R}_d^S(A)$ by $\mathsf{LL}(A)$ for all $A \in \mathcal{A}$ will yield an equivalent argumentation system A′ such that the defeat relation corresponding to A is the same as that corresponding to A′. Furthermore, we have the following result.

Proposition 3. *Let* A *be an argumentation system with elitist preferences satisfying (17), such that* $\mathsf{LL}(A) \subseteq \mathcal{R}_d^S(A)$ *for all* $A \in \mathcal{A}$. *Then* A *is last link representable.*

Proof. We must show both directions of the equivalence in (18). $\Rightarrow$) Let $A, B \in \mathcal{A}$ be arbitrary such that $A \not\prec B$ and assume towards contradiction that there is $r \in \mathsf{LL}(A)$ such that $r \leq q$ for all $q \in \mathsf{LL}(B)$. By (17) and transitivity of $\leq$ it follows that $r \leq s$ for all $s \in \bigcup_{X \in M_S(B)} \mathcal{R}_d^S(X) = \mathcal{R}_d^S(B)$. Hence, $A \leq B$, contradiction. $\Leftarrow$) Let $A, B \in \mathcal{A}$ be arbitrary such that $A = \emptyset$ or $\forall x \in \mathsf{LL}(A) : \exists y \in \mathsf{LL}(B) : x \not\leq y$. If $A = \emptyset$, then clearly $A \not\preceq B$, so assume $\forall x \in \mathsf{LL}(A) : \exists y \in \mathsf{LL}(B) : x \not\leq y$. Assume towards contradiction that $A \leq B$. By (17) there is a witness $r = \mathsf{LastRule}(A_i)$ for some $A_i \in M_S(A)$. Hence, $r \leq q$ for all $q \in \mathsf{LL}(B) \subseteq \mathcal{R}_d^S(B)$, contradicting the assumption. $\qquad\square$

The stronger version of Theorem 2 then follows for strongly structural orderings as a corollary.

Corollary 1. *Assume that* A *is a rule-based argumentation system with conflict contraposition and preferences* $\preceq$, *induced from* $\trianglelefteq$, *which are in turn lifted from a structural preorder* $\leq \subseteq \mathbb{D} \times \mathbb{D}$ *under elitist lifting. Then* A *is consistent if one or more of the following conditions hold.*

    i. $\leq$ *is antisymmetric.*

    ii. $\leq$ *satisfies transitivity of incomparability.*

Proof. Since A is last link representable by Proposition 3, the claim follows from Theorem 3. $\qquad\square$

The structure preference orderings defined by Young and others in [15] are almost strongly structural (since they are strict total orders). Specifically, let $|\mathbb{D}| = \{r \in \mathbb{D} \mid \exists A \in \mathcal{A} : r \in \mathcal{R}_d(A)\}$. Then it is not hard to see that structure preference orderings satisfy the following property:

$$\forall r \in |\mathbb{D}| : \exists A \in \mathcal{A} : \mathsf{LastRule}(A) = r \text{ and } \forall r \in \mathcal{R}_d^S(A) : r \leq \mathsf{LastRule}(A) \qquad (19)$$

That is, for any defeasible rule r that appears in some argument, there is an argument A_r concluding with that defeasible rule, such that A_r also satisfies the condition in (17). Let $|\mathcal{A}| = \{A_r \mid r \in |\mathbb{D}|\}$ be the collection of all such arguments. Then we can show that $|\mathcal{A}|$ provides a normal-form representation of arguments, in the following sense.

Proposition 4. *Let* A *be an argumentation system with preferences satisfying (19). Then for all* $E \in \varepsilon(\mathcal{A}, \mathsf{Def})$ *we have* $|E| \subseteq E$ *where* $|E| = \{A_r \mid \exists X \in E : \mathsf{LastRule}(X) = r\}$.

Proof. Let $E \in \varepsilon(\mathcal{A}, \mathsf{Def})$ and assume towards contradiction that $|E| \not\subseteq E$. Let $A_r \in |E|$ be a minimal argument such that $A_r \notin E$, i.e., such that there is no subargument of A_r that is not in E. Consider arbitrary $X \in E$ such that $\mathsf{LastRule}(X) = r$. By minimality of A_r, every fallible subargument of A_r is in E. Hence, by completeness of E, there must be some $Y \in E$ that defeats A_r on r. But then by (5) Y attacks X as well. Moreover, since A_r is not strictly preferred over Y and $r \le q$ for all $q \in \mathcal{R}_d^S(A_r)$ (by (17)), X is not strictly preferred over Y either (since r is also in $\mathcal{R}_d^S(X)$). It follows that Y defeats X as well, contradicting conflict-freeness of E. $\square$

In view of Proposition 4 and (2), we have $c(E) = c(|E|)$ for all $E \in \varepsilon(\mathcal{A}, \mathsf{Def})$. Hence, $|\mathcal{A}|$ really does provide us with a normal-form representation of arguments under strongly structural preference orderings, meaning also that the consistency results in Theorem 4 and Corollary 1 apply to systems with such orderings. Hence, the consistency results from [15] have been generalised.

7   Conclusion and summary of results

We have studied the relationship between preferences and consistency in structured argumentation. Using the notion of an argumentation signature, we provided a more general version of the reasonableness requirement from [12] (Definition 6). We showed that this requirement ensures consistency for any argumentation system that satisfies certain natural properties, formulated in terms of their signature (Theorem 1). We then went on to consider argumentation systems with preferences, focusing on the case when preferences over arguments are induced by orderings over sets of arguments that are in turn lifted from orderings over defeasible rules. Our signature-based reasonableness requirement was then related to the original notion of reasonableness, which specifically addressed preferences (Proposition 1).

We then considered the elitist lifting principle, returning to the non-strict variant used in the original version of [11], which was later replaced by strict elitist lifting in an erratum. The notion of rationality was introduced, providing a terse and abstract condition on argumentation systems, essentially lifting the idea behind the so-called rationality postulates [3] to the order-theoretic level (Definition 10). Specifically, rationality was shown to imply consistency of (non-strict) elitist lifting (Proposition 2). Using this preliminary result, we then provided sufficient conditions for consistency in terms of more well-known and intuitive order-theoretic conditions, first for weakest link lifting (Theorem 2), then with a stronger version for last link lifting, applicable when only the last defeasible rule of every maximally fallible subargument is taken into account when comparing the strength of arguments (Theorem 3).

Following up on this, we briefly considered a conceptual objection to elitist lifting. We then showed that the disjoint lifting principle that has been proposed to deal with conceptual problems can result in inconsistency even for linear rule-orderings (Example 2). We then considered another recent proposal from the literature, whereby rule-orderings are first reordered to take the structure of arguments into account, before lifting principles are applied [15]. We formulated an abstract condition on orderings that characterise the outcome of such a reordering, giving rise to what we call structural rule-orderings (Eq. 16). We showed that disjoint lifting of a structural rule-ordering is always consistent, thereby explaining and generalising recent work on such liftings [15, 10] (Theorem 4).

However, we also argued that the conceptual problem motivating disjoint lifting is resolved independently (for a different reason) when structural rule-orderings are used. Hence, we suggested that disjoint lifting might be a conceptual dead end. Indeed, elitist lifting of a structural rule-ordering does not suffer from any obvious conceptual shortcomings (that disjoint lifting can resolve). Following up on this, we defined strongly structural rule-orderings and showed that weakest link and last link lifting coincides for such orderings (Proposition 4). In our opinion, this suggests that strongly structural rule-orderings are an interesting special case, where (non-strict and non-disjoint) elitist lifting is conceptually well-behaved and ensures consistency provided the rule-ordering satisfies one of the conditions identified in Theorem 3 (c.f., Theorem 1).

In future work, we would like to address other conceptual objections to elitist lifting, including some addressed in [10] that we have not considered in this article. More generally, we want to consider in more depth the merits and characteristics of strongly structural rule-orderings. It seems to us that such orderings have nice properties and should be investigated further. We also believe the signature-based approach taken in this article is the appropriate way forward, allowing us to discuss the consistency problem abstractly, providing results that are more general and (hopefully) more accessible than results obtained with respect to specific instantiations and constructions.

References

[1] Ofer Arieli and Christian Straßer. Sequent-based logical argumentation. *Argument & Computation*, 6(1):73–99, 2015.

[2] Andrei Bondarenko, Phan Minh Dung, Robert A. Kowalski, and Francesca Toni. An abstract, argumentation-theoretic approach to default reasoning. *Artif. Intell.*, 93:63–101, 1997.

[3] Martin Caminada and Leila Amgoud. On the evaluation of argumentation formalisms. *Artif. Intell.*, 171(5-6):286–310, 2007.

[4] Martin Caminada, Sanjay Modgil, and Nir Oren. Preferences and unrestricted rebut. In Simon Parsons, Nir Oren, Chris Reed, and Federico Cerutti, editors, *Computational Models of Argument - Proceedings of COMMA 2014, Atholl Palace Hotel, Scottish Highlands, UK, September 9-12, 2014*, volume 266 of *Frontiers in Artificial Intelligence and Applications*, pages 209–220. IOS Press, 2014.

[5] Martin W. A. Caminada, Walter Alexandre Carnielli, and Paul E. Dunne. Semi-stable semantics. *J. Log. Comput.*, 22(5):1207–1254, 2012.

[6] Phan Minh Dung. On the acceptability of arguments and its fundamental role in non-monotonic reasoning, logic programming and n-person games. *Artif. Intell.*, 77(2):321–358, 1995.

[7] Phan Minh Dung. An axiomatic analysis of structured argumentation with priorities. *Artif. Intell.*, 231:107–150, 2016.

[8] Phan Minh Dung and Phan Minh Thang. Closure and consistency in logic-associated argumentation. *Journal of Artificial Intelligence Research*, 49:79–109, 2014.

[9] Jesse Heyninck and Christian Straßer. Revisiting unrestricted rebut and preferences in structured argumentation. In *Proceedings of the Twenty-Sixth International Joint Conference on Artificial Intelligence, IJCAI 2017, Melbourne, Australia, August 19-25, 2017*, pages 1088–1092. 2017.

[10] Beishui Liao, Nir Oren, Leendert Van Der Torre, and Serena Villata. Prioritized Norms and Defaults in Formal Argumentation. In *Proceedings of the 13th International Conference on Deontic logic and Normative Systems (DEON 2016)*, Proceedings of the 13th International Conference on Deontic logic and Normative Systems (DEON 2016), Bayreuth, Germany, July 2016. College Publications.

[11] Sanjay Modgil and Henry Prakken. A general account of argumentation with preferences. *Artif. Intell.*, 195:361–397, February 2013.

[12] Henry Prakken. An abstract framework for argumentation with structured arguments. *Argument & Computation*, 1(2):93–124, 2010.

[13] Yining Wu. *Between Argument and Conclusion-Argument-based Approaches to Discussion, Inference and Uncertainty*. University of Luxembourg, 2012. PhD thesis.

[14] Anthony P. Young, Sanjay Modgil, and Odinaldo Rodrigues. Argumentation semantics for prioritised default logic. *CoRR*, abs/1506.08813, 2015.

[15] Anthony P. Young, Sanjay Modgil, and Odinaldo Rodrigues. Prioritised default logic as rational argumentation. In Catholijn M. Jonker, Stacy Marsella, John Thangara-

jah, and Karl Tuyls, editors, *Proceedings of the 2016 International Conference on Autonomous Agents & Multiagent Systems, Singapore, May 9-13, 2016*, pages 626–634. ACM, 2016.

Received 3 July 2017

DEFINING ARGUMENT WEIGHING FUNCTIONS

THOMAS F. GORDON*
Fraunhofer FOKUS, Berlin, Germany

Abstract

Dung designed abstract argumentation frameworks [8] to model attack relations among arguments. However, a common and arguably more typical form of human argumentation, where pros and cons are weighed and balanced to choose among alternative options, cannot be simply and intuitively reduced to attacks. [12] defined a new formal model of structured argument which generalizes Dung abstract argumentation frameworks to provide better support for argument weighing and balancing, enabling cumulative arguments and argument accrual to be handled without causing an exponential blowup in the number of arguments. Dung proposed a pipeline model of argument evaluation for abstract argumentation frameworks, where first all the arguments are evaluated and labeled, at the abstract level, and then, in a subsequent process, the premises and conclusions of the arguments are labeled, at the structured argument level. This pipeline model makes it impossible to make the weight of arguments depend on the labels of their premises. To overcome this problem, in the new model of [12] the weight of arguments and labels of statements can depend on each other, in a mutually recursive manner. The new model is a framework which can be instantiated with a variety of argument weighing functions. In this article, this feature is illustrated by defining a number of argument weighing functions, including: 1) simulating linked and convergent arguments, by making the weight of an argument depend on whether all or some of its premises are labeled **in**, respectively; 2) making the weight of an argument depend on one or more meta-level properties of the argument, such as the date or authority of the scheme instantiated by the argument; 3) modeling a simple form of cumulative argument, by making the weight of an argument depend on the percentage of its **in** premises; 4) making the weight of an argument depend on the percentage of its **in** "factors", from a set of possible factors, where premises represent factors; and, finally 5) making the weight of an argument depend on a weighted sum of the **in** properties of an option, in the style of multi-criteria decision analysis, where premises model properties of an option.

Keywords: artificial intelligence, computational models of argument, multi-criteria decision analysis

*I would like to thank Doug Walton for his many years of support and collaboration, as well as Horst Friedrich of Fraunhofer FOKUS, for his essential and extensive contributions to version 4 of the Carneades software.

1 Introduction

A wide class of human argumentation involves the weighing and balancing of pros and cons to choose among a set of options, including practical reasoning about choosing a course of action, theoretical argumentation both in natural science and the humanities, about which theory is most coherent, factual argumentation about whether or not some claim about an event is sufficiently supported by evidence, as well as reasoning about how best to interpret some abstract ("open-textured") concept in more concrete situations.

The leading computational model of argument, abstract argumentation frameworks [8], was not designed to handle balancing arguments, but rather only to resolve attacks among arguments, especially when attacks are cyclic, i.e. when arguments attack each other, directly or indirectly. Most of the leading models of structured argumentation [3, 16, 25] are defined as preprocessors for an argument evaluator for abstract argumentation frameworks, following a pipeline methodology proposed by Dung in [8, pg 348], in which an abstract argumentation framework is first generated or constructed from domain knowledge and then evaluated to determine which arguments are acceptable. In practice, structured models of argument extend this pipeline with an additional process at the end, for labeling the *statements* (propositions) in the structured model of argument. Typically, a statement is acceptable (in) if and only if it is supported by an acceptable argument.

The linearity of this pipeline presents a problem when one wants to model balancing arguments, since in general the weight of an argument can depend on the acceptability of its premises and, recursively, the acceptability of the premises can depend on the weights of the arguments supporting them. When balancing, pro and con arguments are weighed against each other. An out premise can reduce or strengthen the weight of an argument, without defeating it completely. For an example of reducing the weight of an argument, consider a convergent argument from the testimony of multiple witnesses, where one or more of the witnesses lied. That the failure of a premise can also strengthen an argument, increasing its weight, may seem somewhat surprising or counterintuitive. But consider for example a practical reasoning argument in favor of purchasing a particular car, where it is claimed the car is moderately safe. But if it turns out that the car is actually safer than claimed, the argument in favor of the car is strengthened, *a fortiori*.

In [12], Douglas Walton and I presented a structured argumentation framework with support for both attacks and balancing, which preserves and models the recursivity of the weighing and balancing process, in which the labels of arguments can depend on the labels of statements, and vice versa. The model also formalizes issue-based information systems [14], a leading informal model of argument designed to support practical reasoning and decision-making. Issues have a set of options, statements representing alternative actions or choices, where each option can be supported by arguments. The premises of arguments can, recursively, be called into question by raising issues about them. Issues are resolved by applying proof standards, which are modeled as functions for aggregating the arguments and selecting at most one option. The proof standards make use of argument weighing functions. The framework can be instantiated with any number of weighing functions. Each argument is associated with a weighing function, typically via the argumentation scheme used to construct the argument.

The main aim of this article is to illustrate the concept and use of argument weighing functions in the structured argumentation framework of [12]. Five example weighing functions are defined, for: 1) simulating linked and convergent arguments, by making the weight of an argument depend on whether all or some of its premises are labeled **in**, respectively; 2) making the weight of an argument depend on one or more meta-level properties of the argument, such as the date or authority of the scheme instantiated by the argument; 3) modeling a simple form of cumulative argument, by making the weight of an argument depend on the percentage of its **in** premises; 4) making the weight of an argument depend on the percentage of its **in** "factors", from a set of possible factors, where premises represent factors; and, finally 5) making the weight of an argument depend on a weighted sum of the **in** properties of an option, in the style of multi-criteria decision analysis, where premises model properties of an option.

The rest of this article is organized as follows: the next section presents our formal model of structured argument, with support for argument weighing and balancing. This is followed by a section presenting the five examples of weighing functions, illustrating the formal model. Next there is a section discussing some related work. The articles ends with our conclusions and some ideas for future work.

2 Formalizing Argument Weighing and Balancing

In this section, the formal model of structured argument of [12], supporting argument weighing and balancing, is presented. This section presumes some familiarity with other formalizations of structured argument, in particular ASPIC+ [20]

2.1 Structure

Let $\mathscr{L}$ be a logical language for expressing statements (propositions). As in ASPIC+ [20], this formal model of structured argument is a "framework". It can be instantiated with any logical language.

An *argumentation scheme* is an abstract structure (signature) in this framework providing functions for generating, validating and weighing arguments. The framework can be instantiated with various argumentation scheme structures satisfying this signature. For our purpose here of modeling balancing arguments, only the weighing functions of argumentation schemes are relevant. See Definition 5 for the signature and further details of weighing functions.

Definition 1 (Argument). *An* argument *is a tuple* (s, P, c, u)*, where:*

- *s is the scheme instantiated by the argument*

- *P, the* premises *of the argument, is a finite subset of* $\mathscr{L}$

- *c, a member of* $\mathscr{L}$*, is the* conclusion *of the argument, and*

- *u, a member of* $\mathscr{L}$*, is the* undercutter *of the argument.*

This model of argument closely fits the usual conception of an argument in informal logic and argumentation theory in philosophy [31]. Notice that an argument here, unlike in ASPIC+, is not a complete proof tree, but rather only a single inference step in such a proof tree. Undercutters here are modeled in the same way as in ASPIC+, with a proposition in $\mathscr{L}$ for each undercutter.[1] In practice, these propositions will typically be constructed by applying some predicate to a term naming the argument, such as undercut(a_1). But this is a detail to be worked out when instantiating the framework. Arguments which have undercutter statements as their conclusion are also called undercutters. Notice that the argument includes a reference to the scheme used to construct (or reconstruct) the argument. This will be used to weigh the argument.

Example 1. *Following the tradition of [5], let us use as our running example a practical reasoning task about choosing a car to buy. Let us assume that a domain-dependent argumentation scheme for car buying has been defined, where the premises express the claimed properties of a particular car, one for each of the criteria to be considered, and the weighing function of the scheme computes a weighed sum of the* proven *(not claimed) properties of the car, where the weight assigned to each property by the scheme is chosen to reflect the relative importance of the criterion, relative to the other criterion, in the manner of multi-criteria decision analysis. Here is an example of an argument for a particular auto, applying this scheme:*
Let $a_5 = (s, P, c, u)$ be an argument for buying a Porsche, where:

- *s is a car buying scheme, described in more detail in Section 3.5.*

- *P, the premises, are:*

---

[1] There are three kinds of attacks in ASPIC+: 1) An *undercutter* attacks the applicability of an argument, i.e. the link between the premises and the conclusion of the argument; 2) A *rebuttal* attacks the conclusion of the argument, for example with an argument for a contrary conclusion; and 3) A *premise defeater* attacks a premise of an argument.

1. *type(porsche,sports)*

2. *price(porsche,high)*

3. *safety(porsche,medium)*

4. *speed(porsche,fast)*

- *c, the conclusion, is buy(porsche), and*

- *u, the undercutter, is undercut(a_5)*

Definition 2 (Issue). *An issue is a tuple (O, f), where:*

- *O, the* options *(also called* positions*) of the issue, is a finite subset of $\mathscr{L}$.*

- *f, the* proof standard *of the issue, is a boolean function which evaluates an option. If the function, applied to the option, evaluates to true, the option is said to satisfy the proof standard. See Definition 6.*

Issues are inspired by Issue-Based Information Systems (IBIS) [14]. They extend the concept of a "contrary" in the ASPIC+ model of structured argument, from a binary relation to an n-ary relation. Allowing more than two options is important for two reasons:

1. To allow more than two alternative options in deliberation dialogues and other decision-making contexts.

2. To avoid false dilemmas, by allowing alternatives other than true or false (or yes or no).

A common example used to illustrate false dilemmas is the issue "Have you stopped beating your spouse?" If you answer "yes", this suggests that you had been beating your spouse previously. If you answer "no", this suggests you are still beating your spouse. Thus, it is useful to be able to assert a third option meaning "I have never beat my spouse."

Proof standards of issues are borrowed from the 2007 version of Carneades [11]. Associating proof standards with issues is designed to assure that the same proof standard applies to every position of the issue.

Definition 3 (Argument Graph). *An argument graph is a tuple (S, A, I, R), where:*

- *S, the* statements *of the argument graph, is a finite subset of $\mathscr{L}$.*

- *A, the* assumptions, *is a subset of S, where each statement in A is assumed to be provable or acceptable to the relevant audience without requiring proof.*

- *I, the* issues *of the argument graph, is a finite set of issues, where every position of every issue is a member of S and no s ∈ S is a position of more than one i ∈ I, and*

- *R, the* arguments *of the argument graph, is a finite set of arguments, where all conclusions, premises and undercutters are members of S.*

The restriction requiring each statement to be a position of at most one issue may seem limiting, but this limitation, if it is one, holds also in prior models of structured argument, where implicitly all issues are Boolean, with just two contrary positions, for example P and $\neg P$. That is, every pair of contraries are in effect positions of exactly one issue in these systems. We have generalized issues to allow for more than two positions (options), but have retained the restriction that every statement is a position of at most one issue.

Argument graphs are called graphs for historical reasons. Admittedly this a bit of an abuse of terminology. But every argument graph (S, A, I, R) can be easily mapped to a directed graph (V, E) as follows:

- The vertices, V, of the graph consist of the statements (S), issues (I) and arguments (R) of the argument graph.

- The edges, E, of the graph are constructed by linking arguments in A to their premises, conclusions and undercutters in S, and issues in I to their options in S, in the obvious way.

In most other models of structured argument, argument graphs for structured arguments are not formally defined. In [3], Besnard and Hunter use the term "argument graph" as a synonym for abstract argumentation frameworks. In ASPIC+ arguments are proof trees. Sets of such arguments are often visualized in ASPIC+ presentations as an argument graph, where each argument is a subgraph of the argument graph, but the argument graph per se is not a part of the formal ASPIC+ model.

Example 2. *Figure 1 shows an argument graph for the car buying example, with an argument for buying a Porsche and another argument for buying a Volvo. The labels of the statement nodes, displayed with colors, and arguments, displayed as numbers (weights) on the edges from the arguments to their conclusions, are explained in Section 2.2. The proof standard "PE" used by both issues, means "preponderance of the evidence" and is also defined in Section 2.2. Undercutters are visualized with dashed edges.*

2.2 Semantics

The semantics of argument graphs is defined here in a way inspired by and analogous to the labeling semantics of abstract argumentation frameworks [2], but without mapping argument graphs to abstract argumentation frameworks.

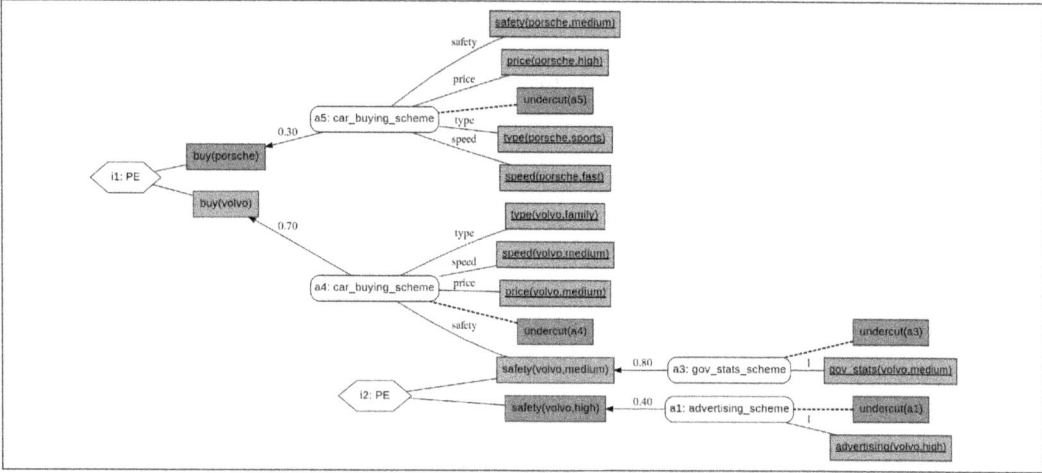

Figure 1: Example Argument Graph

Definition 4 (Labeling). *A labeling is a total function from $\mathscr{L}$ to $\{in, out, undecided\}$.*

Notice that *statements*, not arguments, are labeled **in**, **out**, or **undecided** here, unlike the labeling semantics for abstract argumentation frameworks. Arguments here are labeled by their weights, as described below.

Example 3. *In the argument graph diagram (argument map) shown in Figure 1, the statements shown in green and red are labeled **in** and **out**, respectively. All other statements in $\mathscr{L}$ are, by default, labeled **undecided**. The underlined statements are **in** because they are assumptions. The undercutters are **out** because they are not supported by any arguments and have not been assumed. The* safety(volvo,medium) *position of issue i_2 is **in**, because it is supported by a stronger argument, with a weight of 0.8, than the argument supporting the other position of the issue,* safety(volvo,high), *and issue i_2 is using the preponderance of the evidence proof standard. Finally, the* buy(volvo) *position of issue i_1 is **in**, rather that the* buy(porsche) *position, because the car buying scheme uses a multi-criteria decision analysis weighing function to balance the attributes of the cars.*

Now we are in position to define weighing functions of argumentation schemes more precisely.

Definition 5 (Weighing Function). *A weighing function maps (labeling $\times$ argument graph $\times$ argument) tuples to normalized weights, real numbers in the range of 0.0 to 1.0. Every weighing function must assign the weight of 0.0 to an argument in a labeling if its undercutter is **in** in the labeling.*

Notice that the weight of an argument can depend on:

753

- the labeling

- properties of the argument graph, including but not limited to properties of other arguments about the same issue

- properties of the argument, such as the scheme applied

It is the potential dependence of the weight of an argument on the labeling of statements in the argument graph which makes it unclear how this model could be mapped to the pipelined evaluation model of abstract argumentation frameworks, where the labeling of all statements takes place at the end of the pipeline, after all the arguments have been labeled. Not all weighing functions may be sensible. An interesting project for future work might be to define further rationality constraints for weighing functions, in addition to assuring that undercut arguments weigh 0.0.

Several examples of weighing functions are in Section 3.

Definition 6 (Proof Standard). *A* proof standard *is a mapping from* (*labeling* × *argument graph* × *statement*) *to* {**true**, **false**}. *A statement s* satisfies *a proof standard, f, given a labeling l and argument graph AG, iff $f(l, AG, s) =$ **true**. *Since proof standards are used to justify decisions, a proof standard may allow at most one position of an issue to satisfy the standard.*

Example 4. *The* preponderance of evidence *proof standard can be defined as follows: a position of an issue satisfies the preponderance of evidence standard in an argument graph AG, if and only if there exists an argument in AG for this position (i.e. having this position as its conclusion) which weighs more than every argument in AG for every other position of the same issue, where the weight of an argument, a_i, is derived by applying the weighing function of the argumentation scheme of a_i to (l, AG, a_i).*

Definition 7 (Applicable Argument). *An argument $r \in R$ is* applicable *in a labeling l if and only if:*

- *The undercutter of r is* **in** *or*

- *The undercutter of r is* **out** *in l and every premise of r is not* **undecided** *in l.*

Notice that premises of an argument need not be **in** for the argument to be applicable. Premises that are **out** can weaken or strengthen the argument, without causing it to become inapplicable. Also, somewhat unintuitively, an argument can be applicable even if its undercutter is **in**. Undercut arguments have zero weight. (See Definition 5.)

Example 5. *In the argument graph shown in Figure 1, all of the arguments are applicable, since all of their undercutters are* **out** *and none of their premises are* **undecided***.*

Definition 8 (Supported Statement)**.** *Let AG be the argument graph* (S,A,I,R)*, l be a labeling and s be a statement in S. s is* supported *by AG iff there exists an argument* $r \in R$ *such that*

- *s is the conclusion of r,*

- *r is* applicable *in l, and*

- $w(l,AG,r) > 0.0$*, where w is the weighing function of the scheme of r.*

In other words, a statement is *supported* if it is the conclusion of an applicable argument weighing greater than 0.0. Note that a supported statement is not necessarily labeled **in** in *l*.

Example 6. *In the argument graph shown in Figure 1 the supported statements are*

- `safety(volvo,medium)`,

- `safety(volvo,high)`,

- `buy(volvo)` *and*

- `buy(porsche)`.

Definition 9 (Unsupported Statement)**.** *Let l be a labeling, AG be the argument graph* (S,A,I,R) *and P be the subset of the arguments R having a statement s as their conclusion. s is* unsupported *by the argument graph iff*

- *P is empty or*

- *for every argument* $r \in P$*: r is applicable in l but the weight of r in l is 0.0, i.e.* $w(l,AG,r) = 0.0$*, where w is the weighing function of the scheme of r.*

That is, a statement is *unsupported* if every argument for this statement (i.e. having this statement as its conclusion) is applicable but with a weight of 0.0. Note that supported and unsupported are not duals: A statement can be neither supported nor unsupported.

Definition 10 (Resolvable Issue)**.** *Let* (S,A,I,R) *be an argument graph. An issue* $i \in I$ *is* resolvable *in a labeling l, if for every position p of i: every argument* $r \in R$ *with the conclusion p is applicable in l.*

The basic intuition here is that an issue in an argument graph is ready to be resolved in a labeling, if the labeling provides enough information to evaluate every argument for every position of the issue. It may be that no position of a resolvable issue satisfies its proof standard. Thus being resolvable does not imply that some position of the issue is **in**.

Example 7. *Both issues of the argument graph shown in Figure 1 are resolvable.*

Definition 11 (Conflict Free Labeling). *Let AG be an argument graph (S, A, I, R). A labeling l is* conflict free *with respect to AG iff, for every statement $s \in S$:*

- *if $s \in A$ then $l(s) \neq$ **out***

- *if $s \notin A$ and s is* unsupported *in l then $l(s) \neq$ **in***

- *if s is not a position of some issue $i \in I$ and s is* supported *in l then $l(s) \neq$ **out***

- *if s is a position of some issue $i \in I$ such that i is resolvable in l and s does not satisfy the proof standard of i then $l(s) \neq$ **in***

- *if s is a position of some issue $i \in I$ such that i is resolvable in l and s satisfies the proof standard of i then $l(s) \neq$ **out***

The concept of conflict-freeness here is analogous to conflict-freeness in abstract argumentation frameworks. The purpose is to define constraints which must be satisfied by every labeling of an argument graph. The constraints tell us what the labels may not be, but do not tell us what they must be. Labeling a statement **undecided** is always permitted. So, more precisely, the constraints tells us when a statement may not be **in** or **out**:

- Assumptions may not be **out**.

- An unsupported statement which is not an assumption may not be **in**.

- If a supported statement is not at issue, it may not be **out**.

- If an issue is resolvable and some position of the issue does not satisfy the proof standard of the issue, then the position may not be **in**.

- If an issue is resolvable and some position of the issue satisfies the proof standard of the issue, then the position may not be **out**.

Inspired also by abstract argumentation frameworks, we define the semantics of argument graphs using fix-points of a characteristic function:

Definition 12 (Characteristic Function). *Let AG be an argument graph (S, A, I, R). The characteristic function of the argument graph, f_{AG} : labeling $\rightarrow$ labeling, is defined as follows:*

$f_{AG}(l) =$
   let m be the resulting labeling
   for each $s \in S$:
     if $l(s) \neq$ **undecided** then $m(s) = l(s)$
     else if $s \in A$ then $m(s) =$ **in**
     else if s is unsupported in l
       then $m(s) =$ **out**
     else if s is not a position of some issue and s is supported in l
       then $m(s) =$ **in**
     else if s is a position of some issue $i \in I$ such that
       i is resolvable in l and s does not satisfy the proof standard of i
       then $m(s) =$ **out**
     else if s is a position of some issue $i \in I$ such that
       i is resolvable in l and s satisfies the proof standard of i
       then $m(s) =$ **in**
     else $m(s) = l(s)$

The basic intuition behind this characteristic function is that it is intended to complete a labeling of an argument graph, relabeling some or all **undecided** statements to **in** or **out**, as much as possible in a "single step". The characteristic function can be applied repeatedly (iteratively) until a fix-point is found, i.e. where $f(l) = l$.

Fix-point semantics requires the characteristic function to be monotonic:

Definition 13 (In and Out Statements of a Labeling; Extensions). *Given an argument graph* (S, A, I, R) *and a labeling l, let $i(l)$, called the* extension *of the argument graph in l, denote the subset of S labeled* **in** *in l and $o(l)$ denote the subset of S labeled* **out** *in l.*

Lemma 1 (Preservation of Conflict Freeness by the Characteristic Function). *The characteristic function preserves the conflict freeness of a labeling. That is, for every labeling l, $f_{AG}(l)$ is not conflict free only if l is not conflict free.*

Proof. Let l be a conflict free labeling. Since f_{AG} does not change the labeling of a decided statement, $f_{AG}(l)$ is not conflict free only if some statement s which is **undecided** in l fails to satisfy some constraint in the definition of conflict freeness in $f_{AG}(l)$. Let us assume that $f_{AG}(l)$ is not conflict free and prove the lemma by contradiction. Then there must be some clause in the definition of the characteristic function which can assign an **undecided** statement a label which is not admissible in the definition of conflict freeness. There is no such clause. $\square$

Theorem 1 (Monotonicity of the Characteristic Function). *Let us overload $\subseteq$ to also denote a preorder on labelings, where $l_1 \subseteq l_2$ iff $i(l_1) \subseteq i(l_2)$ and $o(l_1) \subseteq o(l_2)$. The characteristic* *function f is monotonic, preserving this order: for every conflict-free labeling l_1 and l_2, if $l_1 \subseteq l_2$ then $f_{AG}(l_1) \subseteq f_{AG}(l_2)$.*

Proof. Let l_1 and l_2 be conflict free labelings such that $l_1 \subseteq l_2$. Thus, for every statement s, if s is **in** in l_1 then it is also **in** in l_2 and if s is **out** in l_1 then it is also **out** in l_2. The characteristic function never changes the label of an **in** or **out** statement. That is, for every label l, if $l(s) \neq$ **undecided** then $f_{AG}(l)(s) = l(s)$. Thus, if s is **in** in l, then s is **in** in $f_{AG}(l)$ and if s is **out** in l, then s is **out** in $f_{AG}(l)$. Let us assume that $f_{AG}(l_1) \not\subseteq f_{AG}(l_2)$ and prove the theorem by contraction. There must be some **undecided** statement s in l_1 which is not **undecided** in $f_{AG}(l_1)$ and for which $f_{AG}(l_1)(s) \neq f_{AG}(l_2)(s)$. According to the definition of the characteristic function, there are five cases in which an **undecided** statement s can become decided, when s is 1) an assumption; 2) unsupported; 3) supported and not a position of some issue; 4) a position of a resolvable issue, where s satisfies the proof standard of the issue; and 5) a position of a resolvable issue, where s does satisfy the proof standard of the issue. In each case, if $f_{AG}(l_1)(s) \neq f_{AG}(l_2)(s)$, then l_2 is not conflict free, by the lemma which states that the characteristic function preserves conflict freeness, contradicting the assumption that it is conflict free. $\square$

Finally, due to the monotonicity of the characteristic function, various fix-point semantics of argument graphs can be defined, in a way similar but not identical to the semantics of abstract argumentation frameworks:

Definition 14 (Fix-Point Semantics). *Let AG be the argument graph (S,A,I,R). A labeling l is:*

- admissible *if and only if l is conflict-free.*

- complete *if and only if l is admissible and $f_{AG}(l) = l$, i.e. l is a fix-point of f.*

- grounded *if and only if l is complete and minimal, i.e. there does not exist a labeling l' such that $l' \subset l$.*

- preferred *if an only if l is complete and maximal, i.e. there does not exist a complete labeling l' such that $l' \supset l$.*

One difference between this semantics and the semantics of abstract argumentation frameworks is that admissible labellings are the same as conflict free labelings here, but not in abstract argumentation frameworks.

Example 8. *The grounded labeling of the argument graph of the running example is shown in Figure 1. The **in** and **out** labels of statements are shown by filling the boxes of the statements with green and red color, respectively. (No statements are **undecided** in the grounded labeling of this argument graph.)*

The formal model has been fully implemented in Version 4 of the Carneades Argumentation System. Carneades is open source software, published using the MPL 2.0 license.[2] Carneades can be used as a command line program or as a web application. You can try out the web version online using the Carneades server.[3] All of the figures for the examples in this article were generated using this implementation.

3   Example Weighing Functions

Recall that weighing functions (Definition 5) have the signature:

$$\text{labeling} \times \text{argument graph} \times \text{argument} \mapsto 0.0...1.0$$

The argument referred to in the signature is the one being weighed. Recall also that weighing functions are constrained to assign the weight of 0.0 to an argument in a labeling if its undercutter is **in** in the labeling, where a labeling is a mapping from statements in the language, $\mathscr{L}$, to {**in**, **out**, **undecided**}. Again, notice that the weight of an argument may depend on the labels of *any* statements in the argument graph, not just its own premises.

This section presents several useful weighing functions matching this signature. We begin by defining weighing functions for simulating linked and convergent arguments from informal logic in philosophy [31].

3.1   Linked and Convergent Arguments

In informal logic, a *linked* argument has weight only if all of its premises are acceptable (presumptively true) and a *convergent* argument has weight only if some premise is acceptable. That is, linked and convergent arguments work similar to conjunction and disjunction operators in propositional logic.

Let premises(a, g) denote the set of premises and undercutter(a, g) denote the undercutter statement of the argument a in the argument graph g. Now, weighing functions for linked and convergent arguments can be formally defined as follows:

$$\text{linked}(l, g, a) = \begin{cases} 1.0 & l(\text{undercutter}(a, g)) \neq \textbf{in} \wedge \forall p \in \text{premises}(a, g).l(p) = \textbf{in}, \\ 0.0 & \text{otherwise.} \end{cases} \tag{1}$$

$$\text{convergent}(l, g, a) = \begin{cases} 1.0 & l(\text{undercutter}(a, g)) \neq \textbf{in} \wedge \exists p \in \text{premises}(a, g).l(p) = \textbf{in}, \\ 0.0 & \text{otherwise.} \end{cases} \tag{2}$$

---

[2] https://github.com/carneades/carneades-4
[3] http://carneades.fokus.fraunhofer.de/carneades

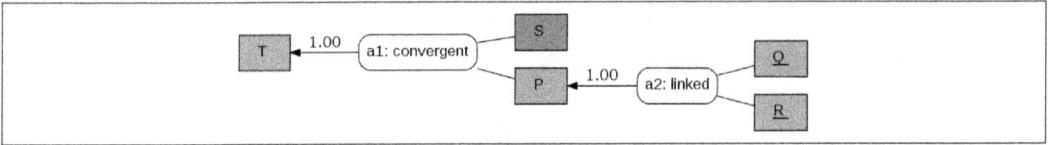

Figure 2: Example of Linked and Convergent Arguments

3.2 Cumulative Arguments

Intuitively, a *cumulative argument* is an argument whose strength increases with the number of its acceptable premises: the greater the number of premises of the argument which are acceptable, the greater the weight of the argument.

Carneades, the Greek skeptic philosopher, gave an early example of a cumulative argument [32], about whether a coiled object in a dark cave is a rope or a snake. He proposed a three-pronged test for deciding whether or not the object is a snake: 1) First, check whether the object looks like a snake; 2) then prod the object with a stick to see if it moves; 3) and finally, jump over the object, again to see if it moves. The greater the number of these tests with positive results, the stronger the presumption that the object is a snake.

Cumulative arguments are special case of argument accrual [27, 19]. Whereas accrual can both strengthen or weaken arguments, cumulation can only strengthen arguments. While there have been efforts to formally model accrual [27, 19], it remains unclear how the weight of a cumulative argument should be affected by the defeat of one or more of its premises. Here, for the sake of illustrating argument weighing functions, we take a straightforward but perhaps too simplistic approach, where the weight of a cumulative argument depends on the proportion of its premises which are acceptable. A more satisfying approach might make use of probability theory [29, 30], but such an approach would of course have the disadvantage of requiring more domain knowledge about probabilities and dependencies.

Here is the formal definition of this simple weighing function for cumulative arguments:

$$\text{cumulative}(l, g, a) = \begin{cases} 0.0 & l(\text{undercutter}(a, g)) = \textbf{in}, \\ r & \text{otherwise.} \end{cases} \tag{3}$$

where

$$r = \frac{|\{p \in \text{premises}(a, g).l(p) = \textbf{in}\}|}{|\text{premises}(a, g)|} \tag{4}$$

Figure 3 shows an argument map for the snake or rope example, with two cumulative arguments, one supporting the conclusion that the object is a rope and the other supporting the conclusion that the object is a snake. Whereas the rope passes all three of the test proposed by Carneades, the Greek philosopher, the snake passes only one of the tests. Thus, the argument for the rope is stronger, with a weight of 1.0, because all three of its premises are **in**, whereas the argument for the snake is weaker, with a weight of only 0.33, because only one of its three premises is **in**.

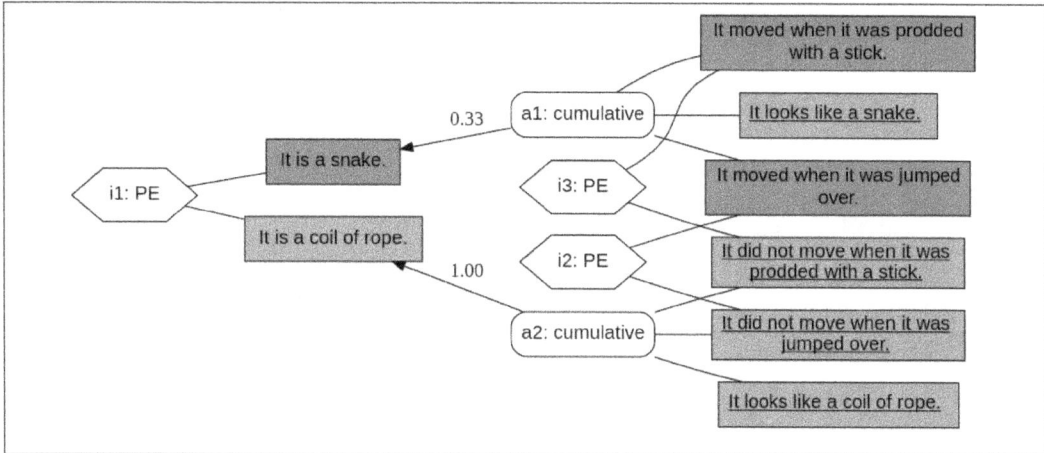

Figure 3: Snake or Rope Example of a Cumulative Argument

3.3 Factorized Arguments

Henry Prakken illustrated the idea of argument accrual with an example about jogging [19]. Some jogger doesn't like to run when it is hot or raining, exclusively, but does like to run when it is both hot and raining. This example illustrates a form of accrual which does not seem to be cumulative, since the conclusion switches from not jogging to jogging, rather than strengthening the argument for not jogging, when a further premise, raining or hot, becomes (presumptively) true.

Figure 4 shows an argument map for the jogging example. What we want is a weighing function which makes jogging **in** when it is both hot and raining, but for not jogging to be **in**, instead, when it is either hot or raining, but not both. There are three arguments in the diagram, one for not going jogging when it is raining (a1), another for not going jogging when it is hot (a2), and a third one for going jogging when it is both hot and raining (a3). If these arguments were weighed using the cumulative weighing function, defined in Section 3.2, then all three arguments would have the same weight, 1.0, because 100% of the premises of all three arguments are **in**. And thus there would no reason for preferring jogging to not jogging and both options would be **out**.

The solution idea is to define a weighing function based on the complete set of *factors* to consider when evaluating each of the options, where each factor is modeled as a statement (proposition). In the jogging example, the set of factors is {hot, raining}. The weight of

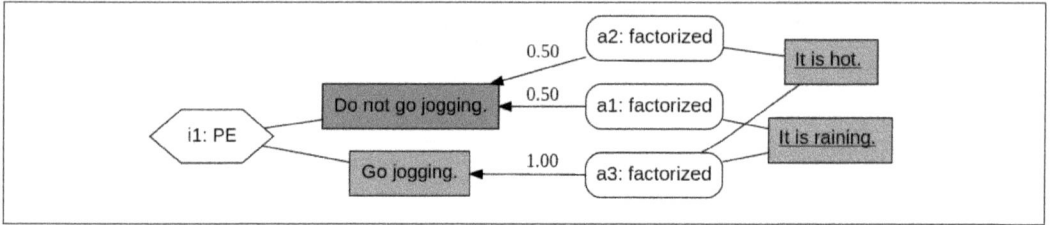

Figure 4: Jogging Example of Factorized Arguments

an argument for some option is then measured as the proportion of the factors which are **in** premises of the argument.

Let factors(a, g) denote the set of factors of the argument a in the argument graph g. Now the factorized weighing function can be formally defined as follows:

$$\text{factorized}(l, g, a) = \begin{cases} 0.0 & l(\text{undercutter}(a, g)) = \textbf{in}, \\ r & \text{otherwise.} \end{cases} \tag{5}$$

where

$$r = \frac{|\{p \in \text{premises}(a, g).l(p) = \textbf{in}\}|}{|\text{factors}(a, g)|} \tag{6}$$

Notice that determining these factors requires traversing the graph from the argument to the issue of the argument and then computing the union of the premises of all the arguments for all of the options of the issue. The issue of an argument is two edges (links) away from the argument in the graph, and the premises of the arguments for each option of the issue are three edges away from the issue in the graph. Thus this method of calculating the weight of an argument cannot be straightforwardly modeled using structured argumentation frameworks such as Abstract Dialectical Frameworks [6], in which the value of a node depends only on its immediate predecessors in a directed graph.

It is possible to reconstruct the jogging example using only cumulative arguments, as shown in Figure 5. The trick is to add further premises to the arguments for not going jogging, for the negation of the other factor; that is, by adding ¬hot as a premise to the argument for not going jogging when it is raining, and adding ¬raining as a further premise to the argument for not going jogging when it is hot. Moreover, we also need to add two issues to the argument graph, for each of the pairs of complements, {raining, ¬raining} and {hot, ¬hot}. We leave it up to the reader to decide which approach to modeling the jogging example is more straightforward and intuitive. Surely the version of the argument graph using the factorized weighing function has fewer nodes, with only half as many premises and one third as many issues. Our purpose here is only to illustrate a variety of possible argument weighing functions.

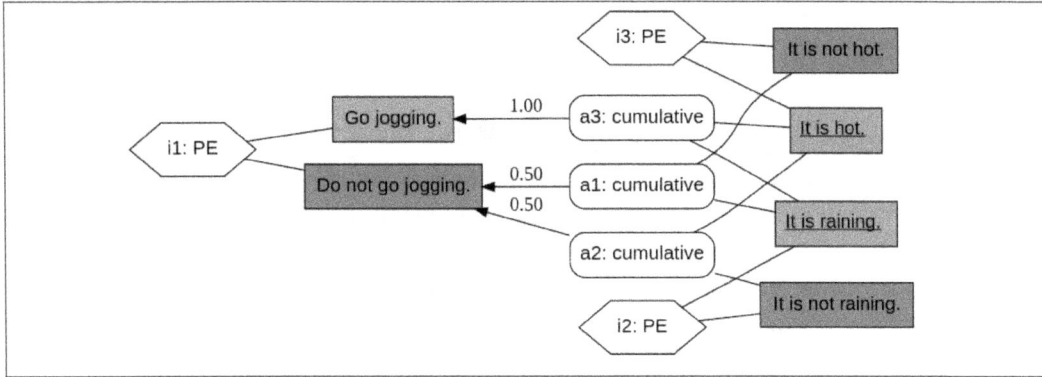

Figure 5: Jogging Example Reconstructed to Use Cumulative Arguments

3.4 Using Metadata to Weigh and Sort Arguments

In the law, the weight of a legal argument can depend on meta-level properties of the argument, or the argumentation scheme used to construct the argument, such as date of enactment of some norm (e.g. rule of law or precedent court case), or the level of authority of the norm in a hierarchical system of authorities (e.g. federal or state law). The law provides principles for resolving these conflicts: *lex superior* gives priority to norms from higher authorities and *lex posterior* gives priority to later norms. (A further principle, *lex specialis*, gives priority to more specific norms.) Moreover, it may be necessary to order these principles, for example to resolve conflicts between earlier federal law and later state law by giving priority to *lex superior* over *lex posterior*.

Prior work on computational models of argument has approached this problem by supporting meta-level argumentation about argument weights or preference orderings [9, 21, 15]. Here we present an alternative approach, using weighing functions to sort arguments based on meta-level properties of the arguments.

Let us model metadata as functions (maps) from property names to values, i.e. functions of the type string $\mapsto$ any. Let metadata(a, g) denote the metadata of the argument a in the argument graph g.

Next, we need a way to order arguments based on their metadata. Let us model this order as a sequence of pairs, where each pair consists of the name of a property and either an implicit order (ascending or descending) or an explicit order (a sequence of property values). The order of the sequence is significant: Properties earlier in the sequence have priority over later properties.

Using this approach, we can model the ordering of legal arguments using the principles of *lex superior* and *lex posterior* as follows. There are two properties: authority and date. The domain of the authority property, i.e. its possible values, is the extensionally defined set {federal, state}. The domain of the date property is, of course, the infinite set of dates. Now we can define the legal ordering, $>_{lex}$, as follows:

$$>_{lex} = [(\text{authority}, [\text{federal}, \text{state}]), (\text{date}, \text{ascending})] \tag{7}$$

This ordering models *lex superior* by giving federal law priority over state law, using an explicit ordering of the two property values, and models *lex posterior* using an implicit, ascending order, giving later dates priority over earlier dates. The ordering also gives *lex superior* priority over *lex posterior*, as desired, since the authority property is ahead of the date property in the sequence.

Now we can use this ordering to sort arguments into equivalence classes, where all arguments in a class have the same priority. That is, no argument in a class is preferred to any other argument in the same class. Let us define a helping function for this purpose, named *sort*, with the following signature:

$$(\text{order}, \text{issue}, \text{argument graph}) \mapsto [2^{\text{argument}}] \tag{8}$$

The sort function maps an order, issue and argument graph into a sequence of sets of arguments. More precisely, its sorts all the arguments for options of the given issue into a sequence of sets of arguments, where the arguments in each set all have the same priority and the arguments in sets later in the sequence have priority over arguments earlier in the sequence.

Now we just need a way to assign weights to these arguments which preserves the ordering. One simple way to do this is to let the weight equal the 1-based index of the set containing the argument, divided by the number of sets in the sequence. This assures that arguments in the last set in the sequence, the strongest arguments, have a weight of 1.0. Conversely, if there are ten sets in the sequence, arguments in the first set of the sequence will have a weight of 0.1. Let $w(a, S)$ denote the weight of an argument a in a sorted sequence S, defined in this way.

$$\text{sorted}(l, g, a) = \begin{cases} 0.0 & l(\text{undercutter}(a, g)) = \mathbf{in} \vee \exists p \in \text{premises}(a, g).l(p) \neq \mathbf{in} \\ r & \text{otherwise.} \end{cases} \tag{9}$$

where
$$r = w(a, \text{sort}(\text{order}(a, g), \text{issue}(a, g), a)) \tag{10}$$

In Equation 10, $\text{order}(a, g)$ denotes the order applicable to the argument a in the argument graph g, and $\text{issue}(a, g)$ denotes the issue of the argument a in g.

Like all weighing functions, by definition, the sorted weighing function assigns a weight of 0.0 to arguments which have been undercut. And, like the linked argument weighing function, it also assigns zero weight to an argument if any of its premise are not **in**.

Figure 6 shows an argument map illustrating the use of the sorted weighing function, using a simple, fairly abstract legal example. The issue is whether a crime has been committed for consuming cannabis. It has been assumed that cannabis has been consumed, so the facts of the case are not at issue. The only issue is whether cannabis consumption is illegal. According to some "earlier federal law", it is illegal, but according to some "later state law", it is not illegal. The same sorting weighing function has been assigned to both of these argumentation schemes and thus, indirectly, to the arguments instantiating these schemes, arguments a_1 and a_2 in the figure. This sorting weighing function gives *lex superior* priority over *lex posterior*. Thus, as we can see in the figure, it is presumptively true that a crime has been committed.

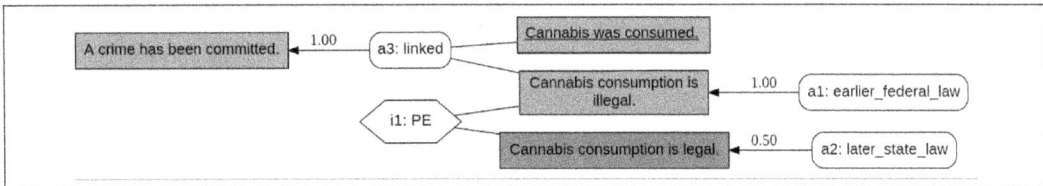

Figure 6: Legal Example of How to Sort Arguments Using Metadata

3.5   Multi-Criteria Decision Analysis

Finally, in our last example of how to define argument weighing functions we show how to use this argumentation framework to realize a form of multi-criteria decision analysis. For this purpose, we reuse the example about choosing a car to buy, shown in Figure 1.

The weighing function of the car buying scheme in this example computes a weighted sum of the proven (not claimed) properties of an option, where the proven properties are represented by the **in** premises of the argument. For example, recall the example argument for buying a Porsche, $a_5 = (s, P, c, u)$, where:

- s is the car buying scheme

- P, the premises, are:

  1. type(porsche,sports)

  2. price(porsche,high)

  3. safety(porsche,medium)

  4. speed(porsche,fast)

- c, the conclusion, is buy(porsche), and

- u, the undercutter, is undercut(a_5)

Each of the premises of the argument represent the value of some property of the Porsche: type, price, safety and speed. The properties are encoded with binary atomic formulas of the form property(object, value). Given an atomic formula of this form, let *property* and *value* be functions for selecting the property and value of the formula, respectively. For example:

$$property(type(porsche, sports)) = type \tag{11}$$

and

$$value(type(porsche, sports)) = sports \tag{12}$$

Given this convention for encoding properties, a multi-criteria decision analysis weighing function can be specified using a structure of the form:

$$([h_1, ..., h_n], [(s_1, w_1, [(q_1, v_1), ..., (q_n, v_n)]), ..., (s_n, w_n, ...)])$$

where each h_i and s_i defines a *hard constraint* and *soft constraint*, respectively.

Each hard constraint, h_i, is a predicate symbol denoting some property of the objects to be compared which *must* be satisfied. To assure this, if a hard constraint of an argument is not **in** in a labeling, the multi-criteria decision analysis weighing function assigns the argument zero weight. For example, when selecting a car to buy, satisfying safety regulations might be a hard constraint. This hard constraint could be modeled with a unary predicate legal. In the argument for buying a Porsche, this would be used by adding a premise legal(porsche) as a hard constraint.

The soft constraints, $s_1, ..., s_n$, are used to compute the weighted sum of the remaining properties, i.e. the properties which do not denote hard constraints. In a soft constraint, w_i is an integer denoting the relative weight (factor), not a percentage, of the property and the (q_i, v_i) assign a numeric value v_i, a real number in the range of 0.0 to 1.0, to the qualitative value (term) q_i of the property. That is, a soft constraint maps the symbolic values of the property to quantities and assigns a relative weight to the constraint.

Let us illustrate this structure for representing weighted sums with the car purchasing example. In this example, the pair of hard and soft constraints is ([], s), where s is

$$\begin{aligned}
[(type, \quad &2, \quad [(sports, 0.0), (sedan, 0.5), (family, 1.0)]), \\
(price, \quad &2, \quad [(low, 1.0), (medium, 0.5), (high, 0.0)]), \\
(speed, \quad &2, \quad [(slow, 0.0), (medium, 1.0), (fast, 0.5)]), \\
(safety, \quad &4, \quad [(low, 0.0), (medium, 0.5), (high, 1.0)])]
\end{aligned} \tag{13}$$

The car purchasing example has no hard constraints, so the first element of the pair is empty, []. There are four soft constraints (type, price, speed, and safety). They are all

766

weighed with the same factor, 2, except for safety, which is deemed twice as important and thus given a factor of 4. The absolute value of the factors is not relevant, just their relation to the other factors. Using factors instead of percentages to model relative weights makes it easier to add or delete properties, without having to reassign the weights or assure they add up to 100. Notice that the value of the speed dimension is non-linear: Medium speed cars are valued higher (1.0) than both slow (0.0) and fast (0.5) cars.

We are now ready to define the weighing function for multi-criteria decision analysis (MCDA). Let $\text{hard}(a, g)$ be the set of premises of the argument a in the argument graph g representing the hard constraints; let $\text{soft}(a, g)$ be the set of soft constraints of a in g; and, for each soft constraint $c_i \in \text{soft}(a, g)$, let $\text{value}(c_i, l)$ denote the quantitative value of the constraint c_i which is **in** in the labeling l (which is not necessarily the value asserted in a premise of the argument being weighed), $\text{factor}(c_i)$ denote the relative weight of the constraint and $\text{sum}(c_i)$ denote the sum of the factors of the constraint.

$$\text{mcda}(l, g, a) = \begin{cases} 0.0 & l(\text{undercutter}(a, g)) = \textbf{in} \vee \exists p \in \text{hard}(a, g).l(p) \neq \textbf{in} \\ w & \text{otherwise.} \end{cases} \tag{14}$$

where

$$w = \sum_{i=1}^{n} \frac{\text{factor}(c_i)}{\text{sum}(c_i)} \times \text{value}(c_i, l) \tag{15}$$

Let us illustrate w by applying it to compute the weighted sum of the soft constraints of the argument for buying a Porsche. The sum of the factors is $2 + 2 + 2 + 4 = 10$. Thus we have

$$\begin{aligned} & 0.0 = 2/10 * 0.0 \quad \text{sports car} \\ + \; & 0.0 = 2/10 * 0.0 \quad \text{high price} \\ + \; & 0.1 = 2/10 * 0.5 \quad \text{fast speed} \\ + \; & 0.2 = 4/10 * 0.5 \quad \text{medium safety} \\ = \; & 0.3 \end{aligned} \tag{16}$$

Since there are no hard constraints and the argument has not been undercut, the weight of the argument, 0.3, is the same as the value of w.

Figure 1 shows an argument map applying this MCDA argument weighing function for the car-buying example. Notice that an issue has been made out of the premise of a_4 stating that Volvos have medium safety, in issue i_2. Had this issue been resolved in favor of the other position of the issue, claiming that Volvos are not merely medium safe, but rather highly safe, then the argument for buying the Volvo, a_4, would have weighed more than it does, 0.7. This illustrates that the failure of a premise can not only weaken an argument, but also strengthen it, *a fortiori*. The example also illustrates how the weight of an argument, a_4, can depend on the label of a statement, safety(volvo,medium), which in turn (recursively) depends on weights of other arguments, a_1 and a_3, which is not possible in a straightforward way using Dung's pipeline model of abstract argument evaluation. (The weights of arguments a_1 and a_3 have been assumed in the example. The weighing functions used to derive these weights are unspecified.)

4 Related Work

The formal model of structured argument has been clearly inspired by Dung's work [8], even if it does not follow his recommended pipeline methodology. Rather, Dung's approach, in particular its use of fix-point semantics, has been used as a model and adapted to the purpose of handling balancing arguments, in addition to attack relations among arguments.

Another source of inspiration for this work was ASPIC+ [20]. All three kinds of attack relations supported by ASPIC+ (premise attacks, rebuttals and undercutters) are also supported in this model. A question for future work is whether ASPIC+ can be simulated using this new model. The new model appears to be both simpler and more expressive than ASPIC+, considered as a whole, despite some elements of the new model being more complex.

Abstract Dialectical Frameworks (ADFs) [6], may seem like a suitable foundation for the new model, since they provide a convenient platform for defining a wide variety of graph-based formalisms. However, ADFs evaluate and label nodes using functions attached to nodes which depend only upon the parents of the nodes, i.e. the immediate predecessors of the node in the directed graph. This does not appear to be general enough for the purposes of the new model, as can be seen in the factorized and MCDA weighing functions, illustrated with the jogging and car buying examples.

In his PhD thesis [26], Tom van der Weide developed a comprehensive formal model of decision-making, based on computational models of argument, covering meta-level reasoning about attributes, values and utility functions, as well as deliberation procedures. However, the thesis did not break new ground with respect to argument accrual, cumulative arguments or methods for weighing arguments. Rather, van der Weide applied Prakken's approach to handling accrual [19] which is computationally and epistemologically less than ideal, due to its causing an exponential blow-up in the number of arguments.

Finally, of course the model is derived from Gordon and Walton's own prior work on structured argumentation with Prakken [11] and preserves all of its features, including its support for variable proof standards and its support for modeling the two kinds of critical questions of argumentation schemes, using assumptions and exceptions. However the new model is simpler and more general in several ways. The main additional complexity in the new model is its introduction of a third node type for issues, in addition to statements and arguments. However this complexity is justified by enabling the new model to provide better support for decision-making, in the style of issue-based information systems.

One important advantage of the new formal model is that argument graphs are now much closer to the conceptual model underlying the argument diagrams typically used in informal logic textbooks, such as [31]. This conceptual model underlies several argument mapping tools, including Araucaria [23], and is also the basis for the Argument Interchange Format (AIF) [22]. Version 4 of Carneades, based on the new model presented here, can import and evaluate AIF files.

Let us now turn to prior work on formal models of argument weights. Qualitative Value Logic [5], which first introduced the car-buying example, extended Brewka's Preferred Subtheories approach to nonmonotonic logic [4], with an evaluation function for weighing defaults and showed how to use this logic to support multi-criteria decision making. The evaluation function assigned weights to the default rules and could not depend, recursively, on the set of propositions entailed by the default theory.

The Zeno Argumentation Framework [10] presented a formal model of structured argument for supporting practical reasoning, inspired by issue-based information systems [14]. The model defined dialectical graphs, consisting of issues, propositions (called "positions") and pro and con arguments, where an issue is a set of propositions, representing options, and a set of qualitative constraints on the weights of propositions. Essentially, arguments were weighed and ordered by solving these constraints. Zeno was more limited than the model presented here in a couple of ways: 1) Every argument was limited to having exactly one premise, so the weight of an argument determined the weight of its premise; and 2) dialectical graphs were not permitted to contain cycles. One the other hand, Zeno supported meta-level argumentation about the constraints, which would be comparable to supporting arguments about the weighing functions, which is not supported by the model presented here. This additional expressiveness of Zeno was enabled by the restriction to acyclic dialectical graphs.

Reason-Based Logic [13, 28] resolved conflicts among arguments, called reasons, by accruing and weighing sets of arguments for each option. The model did not include a formalization of weighing functions, but rather assumed the sets of arguments could be weighed in some fashion to produce a priority relation provided as input to the argument evaluation process. As noted by Siekmann [24], Reason-Based Logic does not provide a way to make weights of arguments depend on the evaluation of arguments about the premises of the arguments to be weighed. We agree with Siekmann [24, p. 213] that, in general, "arguments to be balanced are not only the objects of, but also reasons guiding, the balancing process." The model of argument weighing and balancing presented here provides this capability.

Amgound and Prade [1] developed a method for using abstract argumentation frameworks to support decision-making. This work extends a Dung abstract argumentation framework with a set of options, modeling the set of alternatives under consideration, and proposes a two-step decision procedure: 1) pro and con arguments for the options are first evaluated using Dung abstract argumentation frameworks; and, 2) the options are then compared, pair-wise, and ordered, using various "decision principles". These decision principles come closest to the concept of weighing functions presented here. One such principle prefers the option with the greatest number of pro arguments. Another principle prefers the option whose con arguments are all weaker than at least one con argument of the other option. Further principles take into consideration both pro and con arguments. Finally, multi-criteria decision analysis is simulated using aggregation functions for computing a weighted sum of pro and con arguments. Because [1] applies Dung's pipeline model of argument evaluation, it does not handle decision problems in which the acceptability of an argument for some option depends on applying argumentation, recursively, to first resolve an issue about some premise.

Ouerdane, *et al.*, [18] presents a fairly elaborate argumentation-based process model of multi-criteria decision-making. The model applies multi-criteria decision analysis to evaluate and compare options as well a formal model of structured argument for reasoning about the properties of the options. Conflicts among arguments are resolved using meta-level priority rules, which serve as weighing functions. Thus, this work has much in common with [1], which they cite, in that argumentation and decision-making are separate tasks of a procedural model in both. Our work, on the other hand, aims to integrate multi-criteria decision analysis into a formal model of structured argumentation.

Finally, Müller and Hunter [17] developed another decision-making model integrating argumentation and multi-criteria decision analysis. The model is also similar in spirit to [1], in that it models argumentation about the properties of options and the evaluation and comparison of the options, using multi-criteria decision analysis, separately. Müller and Hunter use a simplified version of ASPIC+ [20], without undercutters[4], to generate options and properties of options, using a rule base, as well as to evaluate the resulting arguments, using grounded semantics. The results of the argumentation process are then fed into a decision analysis process to compare the options. The options are weighed and ordered by the number of goals they satisfy in the grounded extension. Thus, this is another pipeline model where argumentation precedes decision analysis. The results of the decision analysis can have no impact on the evaluation of the arguments and thus the arguments may not raise any practical issues which need to be resolved via decision analysis. (In real-life practical reasoning, a decision may depend on other decisions. For example, when planning a vacation, the choice of hotel can depend on the choice of location.) Moreover, the system uses a fixed weighing procedure for comparing options, rather than providing a framework which can be instantiated with various weighing functions, as in our system.

---

[4]Caution: "Hunter calls attacks on premises in ASPIC+ undercutters.

5 Conclusion

This article illustrated the concept of weighing functions in a formal model of structured argument with support for balancing arguments. Five example weighing functions were presented, for 1) simulating linked and convergent arguments, by making the weight of an argument depend on whether all or some of its premises are labeled **in**, respectively; 2) making the weight of an argument depend on one or more meta-level properties of the argument, such as the date or authority of the scheme instantiated by the argument; 3) modeling a simple form of cumulative argument, by making the weight of an argument depend on the percentage of its **in** premises; 4) making the weight of an argument depend on the percentage of its **in** "factors", from a set of possible factors, where premises represent factors; and, finally 5) making the weight of an argument depend on a weighted sum of the **in** properties of an option, in the style of multi-criteria decision analysis, where premises model properties of an option.

The model can handle cumulative arguments [32] and argument accrual [19] without causing an exponential blow-up in the number of arguments. While the model does not map structured arguments to abstract arguments, it is inspired by the fix-point semantics of abstract argumentation frameworks and uses comparable methods to handle and resolve cycles in argument graphs. The formal model has been fully implemented in Version 4 of the Carneades argumentation system, for grounded semantics. All of the figures for the examples in this article were generated using this new version of Carneades.

New in this article is the proof of the monotonicity of the characteristic function of argument graphs, which was left as a conjecture in [12].

Some ideas for future work include formally investigating relationships between this model and other models of structured argument, in particular ASPIC+. Another idea would be to investigate whether or not Caminada's rationality postulates for structured argumentation [7], e.g. closure, direct consistency and indirect consistency, are meaningful in the context of this model and, if so, whether they are satisfied.

The model of structured argument presented in [12] provides a framework which can be instantiated with arbitrary argument weighing functions, but not all weighing functions may be sensible or rational. This paper illustrated how the framework can be instantiated with a variety of argument weighing functions for various purposes, but it remains for future work to develop rationality constraints for weighing functions or to formally investigate the properties of various weighing functions. The rationality constraints may depend on the task or dialogue type. Moreover, the current model allows different argument weighing functions to be applied within the context of a single issue, without restrictions. This seems too unconstrained. Perhaps a concept of *issue schemes* is required, comparable to argumentation schemes, for restricting the allowed schemes of arguments put forward about options of each kind of issue, and thus indirectly restricting the argument weighing functions which may be used in arguments for each kind of issue.

References

[1] Leila Amgoud and Henri Prade. Using arguments for making and explaining decisions. *Artificial Intelligence*, 173(3):413–436, 2009.

[2] Pietro Baroni, Martin Caminada, and Massimiliano Giacomin. An introduction to argumentation semantics. *The Knowledge Engineering Review*, 26(4):365–410, 2011.

[3] Philppe Besnard and Anthony Hunter. Constructing argument graphs with deductive arguments: A tutorial. *Argument and Computation*, 5(1):5–30, 2014.

[4] Gerhard Brewka. Preferred Subtheories: An Extended Logical Framework for Default Reasoning. In *Proceedings of the Eleventh International Joint Conference on Artificial Intelligence*, pages 1043–1048. 1989.

[5] Gerhard Brewka and Thomas F Gordon. How to buy a Porsche: An approach to defeasible decision making. In *Working Notes of the AAAI-94 Workshop on Computational Dialectics*, pages 28–38, Seattle, Washington, 1994.

[6] Gerhard Brewka and Stefan Woltran. Abstract Dialectical Frameworks. In *Proceedings of the Twelfth International Conference on the Principles of Knowledge Representation and Reasoning*, pages 102–111. AAAI Press, 2010.

[7] Martin Caminada and Leila Amgoud. On the evaluation of argumentation formalisms. *Artificial Intelligence*, 171(5-6):286–310, April 2007.

[8] Phan Minh Dung. On the acceptability of arguments and its fundamental role in nonmonotonic reasoning, logic programming and n-person games. *Artificial Intelligence*, 77(2):321–357, 1995.

[9] Thomas F Gordon. *The Pleadings Game; An Artificial Intelligence Model of Procedural Justice*. Kluwer Academic Publishers, Dordrecht; Boston, 1995.

[10] Thomas F Gordon and Nikos Karacapilidis. The Zeno argumentation framework. In *Proceedings of the Sixth International Conference on Artificial Intelligence and Law*, pages 10–18, Melbourne, Australia, 1997. ACM Press.

[11] Thomas F. Gordon, Henry Prakken, and Douglas Walton. The Carneades model of argument and burden of proof. *Artificial Intelligence*, 171(10-11):875–896, 2007.

[12] Thomas F. Gordon and Douglas Walton. Formalizing balancing arguments. In *Proceeding of the 2016 conference on Computational Models of Argument (COMMA 2016)*, pages 327–338. IOS Press, 2016.

[13] Jaap Hage and Bart Verheij. Reason-Based Logic: a Logic for Reasoning with Rules and Reasons. *Information & Communications Technology Law*, 3(2-3):171–209, 1994.

[14] Werner Kunz and Horst W.J. Rittel. Issues as elements of information systems. Technical report, Institut für Grundlagen der Planung, Universität Stuttgart, 1970.

[15] Sanjay Modgil and Henry Prakken. A general account of argumentation with preferences. *Artificial Intelligence*, 195:361–397, 2013.

[16] Sanjay Modgil and Henry Prakken. The ASPIC+ framework for structured argumentation: A tutorial. *Argument and Computation*, 5(1):31–62, 2014.

[17] Jann Müller and Anthony Hunter. An argumentation-based approach for decision making. In *Tools with Artificial Intelligence (ICTAI), 2012 IEEE 24th International Conference on*, volume 1, pages 564–571. IEEE, 2012.

[18] Wassila Ouerdane, Yannis Dimopoulos, Konstantinos Liapis, and Pavlos Moraitis. Towards automating decision aiding through argumentation. *Journal of Multi-Criteria Decision Analysis*, 18(5-6):289–309, 2011.

[19] Henry Prakken. A Study of Accrual of Arguments, with Applications to Evidential Reasoning. In *Proceedings of the Tenth International Conference on Artificial Intelligence and Law*, pages 85–94, New York, 2005. ACM Press.

[20] Henry Prakken. An abstract framework for argumentation with structured arguments. *Argument & Computation*, 1:93–124, 2010.

[21] Henry Prakken and Giovanni Sartor. A Dialectical Model of Assessing Conflicting Argument in Legal Reasoning. *Artificial Intelligence and Law*, 4(3-4):331–368, 1996.

[22] I. Rahwan and Chris Reed. The Argument Interchange Format. In I. Rahwan and Chris Reed, editors, *Argumentation in Artificial Intelligence*. Springer, 2009.

[23] Chris A Reed and Glenn W A Rowe. Araucaria: Software for argument analysis, diagramming and representation. *International Journal of AI Tools*, 13(4):961–980, 2004.

[24] Jan-R. Sieckmann. Why non-monotonic logic is inadequate to represent balancing arguments. *Artificial Intelligence and Law*, pages 211–219, 2003.

[25] Francesca Toni. A tutorial on assumption-based argumentation. *Argument and Computation*, 5(1):89–117, 2014.

[26] Tom van der Weide. *Arguing to Motivate Decisions*. Dutch Research School for Information and Knowledge Systems, 2011.

[27] Bart Verheij. Accrual of arguments in defeasible argumentation. In *Proceedings of the Second Dutch/German Workshop on Nonmonotonic Reasoning*, pages 217–224, 1995.

[28] Bart Verheij. *Rules, Reasons, Arguments. Formal Studies of Argumentation and Defeat*. Ph.d., Universiteit Maastricht, 1996.

[29] Bart Verheij. Correct grounded reasoning with presumptive arguments. In *European Conference on Logics in Artificial Intelligence*, pages 481–496. Springer, 2016.

[30] Bart Verheij. Proof with and without probabilities correct – evidential reasoning with presumptive arguments, coherent hypotheses and degrees of uncertainty. *Artificial Intelligence and Law*, 25(1):127–154, 2017.

[31] Douglas Walton. *Fundamentals of Critical Argumentation*. Cambridge University Press, Cambridge, UK, 2006.

[32] Douglas N. Walton, Christopher W. Tindale, and Thomas F. Gordon. Applying recent argumentation methods to some ancient examples of plausible reasoning. *Argumentation*, 28(1):85–119, 2014.

Received 16 May 2017

www.ingramcontent.com/pod-product-compliance
Lightning Source LLC
Chambersburg PA
CBHW081151090426
42736CB00017B/3264